Spanish Air Force Aircraft

1939–2021

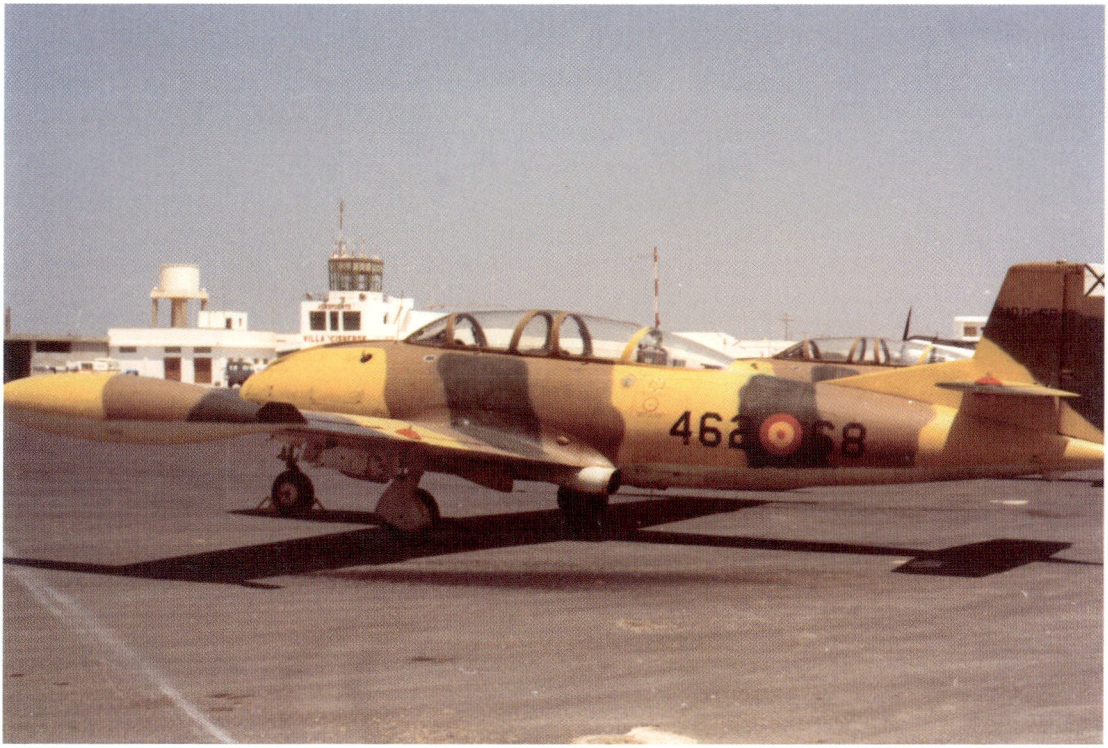

EDUARDO MANUEL GIL MARTÍNEZ
& JUAN ARRÁEZ CERDÁ

KEY Books

AIR FORCES SERIES, VOLUME 3

Front cover image: A Eurofighter from Ala 14 flying over Spain. Curiously, the first Eurofighters deployed to Ala 14 went to 142 Escuadrón, and then to 141 Escuadrón. From their air base in Los Llanos, they can survey a great area of Spain and the Mediterranean Sea.

Back cover image: A C-101 belonging to the Patrulla Águila (Eagle Patrol) showing its beautiful livery. This Spanish-made jet trainer was a milestone for the Spanish aeronautical industries and still soldiers on within the EdA as a trainer and an aerobatic aircraft.

Title page image: Two 'Saetas' from 462 Escuadrón parked at Villa Cisneros Airport (now Dakhla Airport) in Spanish Sahara in 1975. This was the last year that Spain held its Western Africa territories. Thanks to the high performance of the aircraft, the 'Saeta' could land at first class airports and airstrips.

Contents page image: A line-up of F-86s from 111 Escuadrón, waiting for the next mission. These aircraft were an important 'step' for the EdA, in spite of their weary status when they arrived in Spain.

A Solete, mi vida.
A mis padres, Salud y Eduardo.
A Caco, Iñigo, Ibón y June.
A Merce y Ricardo.
A mis abuelos Mercedes, Salud, Mami, Manuel y Juan.

A mi estupendo amigo Blas Vicente Marco, compañero en muchas 'aventuras aeronáuticas' llevadas a feliz término, esperando le guste este libro eminentemente fotográfico sobre lo que fue y es nuestro amado Ejército del Aire.

Acknowledgements
Special thanks to Marisol García Gómez, Ricardo Ramallo Gil, Yevgeniya Akinshina (Airbus Defence, AD), José Gascó (AD), Pilatus Aircraft Ltd, Ejército del Aire, Mª José Rossell (Oficina de Comunicación del EdA) and Pablo Rada.

Almost all pictures are from the archive of Juan Arráez, and those he took himself. Those that are the exception to this have been attributed in the caption.

Published by Key Books
An imprint of Key Publishing Ltd
PO Box 100
Stamford
Lincs PE19 1XQ

www.keypublishing.com

The right of Eduardo Manuel Gil Martínez and Juan Arráez Cerdá to be identified as the authors of this book has been asserted in accordance with the Copyright, Designs and Patents Act 1988 Sections 77 and 78.

Copyright © Eduardo Manuel Gil Martínez and Juan Arráez Cerdá, 2021

ISBN 978 1 80282 034 8

All rights reserved. Reproduction in whole or in part in any form whatsoever or by any means is strictly prohibited without the prior permission of the Publisher.

Typeset by SJmagic DESIGN SERVICES, India.

Contents

Foreword 4

Chapter 1 From Spanish Civil War to 1953: The Reorganisation 5

Chapter 2 From 1953 to 1965: American Aid and Spanish Aircraft Industries Impulse 24

Chapter 3 From 1965 to 1985: Improving the Breed..39

Chapter 4 From 1985 to 2000: Sharpening the Claws...63

Chapter 5 From 2001 until Today: The Most Powerful EdA of all Time? 73

Appendices 82

Bibliography 96

Foreword

Dear reader,

The book you have in your hands has come to life in order to show you a pictorial history of the Spanish Ejército del Aire (EdA, Spanish Air Force) and its aircraft, from its formation until the present day. The pictures are distributed across six chapters, although the number of images sadly declines as the book continues. This is because, while the first years of the EdA were characterised by the large volume of aircraft types that were produced, as we move towards the present, the number of aircraft types produced has reduced significantly. We have primarily used pictures from the files of Juan Arraez to show the aircraft that have flown within the EdA from 1939 until early 2021.

We want this book to be a tribute to all the people who have worked with these aircraft, and who overcame a variety of challenges and difficulties to provide Spain with a modern and highly effective air force.

Before reading this book, it is important to understand the EdA's organisation. Of course, it has changed several times since the end of the World War Two, but essentially is important to know the following:

The basic unit within the EdA is called Ala (Wing). Each Ala has between one and three Escuadrones (Squadrons), but, typically, there are two Escuadrones. When there is a third Escuadrón, it is usually intended as a Conversion Unit for the training of new pilots on new aircraft. Most of the time, the Escuadrón's number begins with the Ala's number, so, for example, within Ala 11, there are three Escuadrones numbered 111, 112 and 113, and within Ala 14 there are Escuadrones 141 and 142.

On occasion, there has been another unit similar to the Ala, which can be joined with another unit: this is the Grupo (Group). For example, in 1989, Ala 15 was renamed as Grupo 15 and was attached to Ala 31, together with Grupo 31.

The number of aircraft in each Escuadrón has varied during the last 75 years, but, on average, it can be between 18 and 24 aircraft. Therefore, the number of aircraft in one Ala can be between 36 and 48. However, the numbers can differ between fighter wings and transport wings. As an example, Ala 11 (a fighter wing) currently has 38 aircraft (Eurofighter EF2000) divided between two Escuadrones: 111 and 113; and Ala 31 (a transport wing) is only intended to have 14 A 400Ms in service.

However, an Escuadrón has not always belonged to an Ala. Escuadrón 104 (F-104G) was an independent unit from November 1967 to May 1972, although, it was at last folded into Ala 12, while retaining the same designation.

Along with Alas and Escuadrones, there is another minor unit called Escuadrilla. Usually, it did not belong to a major unit and was independent.

We hope you enjoy the superb pictures.

Chapter 1

From Spanish Civil War to 1953: The Reorganisation

After the bloody Spanish Civil War (SCW) that took place between 1936 and 1939, Spain was exhausted when it declared the end of the war on 1 April. However, because of the war years, it had accumulated a large amount of very heterogeneous war material.

Shortly after the end of the SCW, the Minitsry of Air was created on 8 August 1939, with General Juan Yagüe, a Legion veteran and Germanophile, elected as minister. On 7 October 1939, by government law, the EdA was born. Initially, all the aircraft in its possession were veterans of the war and, according to the report sent on 9 February 1940 by the EdA general director, it had 1,148 aircraft of 95 different types.

Generalissimo Francisco Franco, head of the Spanish State, approved the regional redistribution of the EdA on 15 July 1939, which divided the Spanish territories into five 'Regions' and three 'Zones'. Numbered from one to five, the former were called the Central, Strait, Levante, Cantabrian and Pyrenean Regions, and the latter were the Canary and Western Africa Air Zone (which included the Canary Islands, Juby Cape, Ifni and Río de Oro), the Balearic Air Zone and the Moroccan Air Zone.

During the period of overlap between the SCW and World War Two, it was evident of the aircraft heterogeneity in the EdA and, in 1940, the aircraft were distributed to the newly established Regions and Zones and their Regimiento (regiments): 1ª Región Aérea (21 Regimiento, 31 Regimiento, Regimiento Mixto), 2ª Región Aérea (11 Regimiento, 12 Regimiento, 22 Regimiento), 3ª Región Aérea (13 Regimiento, 32 Regimiento), 4ª Región Aérea (14 Regimiento, 15 Regimiento, 23 Regimiento), 5ª Región Aérea (16 Regimiento), Fuerzas Aéreas del Protectorado de Marruecos (Regimiento Mixto nº 2), Fuerzas Aéreas de Baleares (Regimiento Mixto nº 3) and Fuerzas Aéreas del Atlántico.

During World War Two, several changes affected the distribution of aircraft to these Spanish territories. The impact of attrition and the lack of spare parts caused by the war meant that many of the EdA's aircraft had become obsolete and, while the EdA was still operating pre-World War Two aircraft, the rest of the world had progressed to jets.

The first task entrusted to the EdA was to collect, and return to flight-ready condition, all the aircraft abandoned by the Spanish Republican Army. This task could not be fulfilled immediately, and it had to wait until the first half of 1940, when some of the Republican aircraft that had been flown to France could be recovered.

A very nice photograph of a Fiat CR.32. Note the Spanish Aviation tricolour badge on the wing, the St Andrew's Cross on the tail and black circle in the fuselage. Although it eventually became obsolete, the CR.32 was the fighter that the EdA utilised in great numbers. Its name in the EdA was C.1, as it was the first fighter type.

Polikapov I-16 'Rata' is photographed in flight. The I-16 was deployed to the Son San Juan AB, Mallorca, to ensure the protection of the Baleares Islands, but it did not last long, being replaced by the Fiat CR.32. Fifty-two I-16s flew for the EdA, first at 8 Grupo in the Balearic Islands and then with 26 Grupo at Tablada. The I-16 ended its operational life at Morón de la Frontera AB, where the last one was retired in September 1953.

The Fiat G.50 fighters were deployed to 2 Escuadrón of 27 Grupo, based in Nador AB, Melilla. In this picture, we can see in a G.50 flying over Nador, adorned with the 2 Escuadrón badge, a white greyhound jumping down inside a white circle. These aircraft suffered various problems with landing gear during their service in North Africa, so they flew fewer hours than the Heinkel He 112, which was also stationed in Nador.

In January 1940, the number of EdA aircraft, according to type, was as follows:

Fighters – 124 Fiat CR.32s, 20 Polikarpov I-15bises, 41 Messerschmitt Bf 109s, 11 Polikarpov I-16s, Fiat G.50 and 20 Heinkel He 112s. This gave a total of 216 aircraft, however, only 161 were operational. In May 1943, 15 Bf 109 F2/F4 fighters, previously used by Luftwaffe units, were purchased from Germany. It is likely that there were ten F4 and five F2 fighters.

Bombers – Ten Fiat BR.20s, 62 Savoia-Marchetti S.79s, 12 Henschel Hs 123s, 18 Tupolev SB-2s, 52 Heinkel He 111s, 15 Savoia S.81s, and 11 Junkers Ju 52s. This gave a total of 180 aircraft, however, only 136 were operational. At the end of November 1943, the EdA purchased ten Ju 88 A-4s from Germany. Furthermore, more Ju 88s – about 18 of several types – were acquired thanks to the internment and requisitioning of these aircraft after the completion of combat missions on Spanish lands.

Assault or Attack – 44 Heinkel He 51s and more than 50 Polikarpov I-15s. This gave a total of around 94 aircraft, all of which were in service.

Seaplanes (in 1943) – Eight Dornier Do J Wals, one Heinkel He 59, four Heinkel He 60s, one Cant Z.506 Airone, and 12 Heinkel He 114s. This gave a total of 26 aircraft, however, only 21 were operational. The last Wal was retired on 15 September 1950. Following the acquisition of 12 Dornier Do 24T-3s, the EdA's capabilities in rescue missions improved significantly. As a curiosity, Spain had one Consolidated Catalina OA-10 (ex-USAAF), which was interned in Spain in July 1943 and acquired from its owner one year later.

Reconnaissance – 18 Heinkel He 45s, eight Dornier Do 17s, 13 Caproni Ca.310s, 34 Polikarpov R-Zs, and five Henschel Hs 126. This gave a total of 78 aircraft, of which only 49 were operational. From 1942, there was an addition of five ex-Republican Grumman GE-23 Delfins, which had been flown to Oran at the end of the SCW and had finally been returned to Spain, as well as four more that had been abandoned at Catalonian aerodromes.

Transport – Four/five Ju 52s, three Messerschmitt Bf 108s, and one Fokker F-XII. This gave a total of eight/nine aircraft, all of which were in service.

Aviation Schools – 11 S.81s, seven Caproni AP.1s, ten Breda Ba.65s, and one Heinkel He 70. This gave a total of 22 aircraft, all of which were in service. The following year, there was an addition of ten CR.32s and three González-Gil Pazó GP-1 aircraft. From 1941 onwards, the Spanish company CASA (Construcciones Aeronáuticas Sociedad Anónima) began to deliver several trainer aircraft, manufactured under the license of a German patent. The first to arrive were 530 Bücker Bü 131s, 25 Bücker Bü 133s and 25 Gotha Go 145s. Although the first Bü 131 aircraft had technically arrived in Spain between 1936 and 1937, and the Go 145 in 1938, they had been externally acquired from Germany.

Meteorological Squad – Two Ju 52s. Both of these aircraft were in service. Three He 111Js joined before February 1941, and at least one He 111H5 was added in December 1943.

At the end of the Spanish Civil War in 1939, the EdA decided to start licence-building of several aircraft types, primarily of German design. Often, when aircraft were built under licence, they received a different name, designed to distinguish those built internally from those acquired externally. CASA received orders to manufacture the following: Bücker Bü 131, which became CASA C-1131; Bücker Bü 133, which became CASA C-1133; Gotha Go 145, which became CASA C-1145; He 111H, which became CASA C-2.111; and Ju 52, which became CASA C-352. Occasionally, the patents would also come from Italian designers. For example, another Spanish manufacturer, Hispano Aviación, began the licensed production of the CR.32, known in Spain as HA-132L.

One of the main aircraft to be license-built in Spain was the Bf 109 G-2. Indeed, 25 engineless fuselages of this fighter were purchased from Germany and arrived in Spain in 1944. However, the lack of the DB engines required it to be replaced with the 12Z-89 French-designed engine. Although, at the beginning, this aircraft was named Me 109 J, the denomination changed to HA-1109 J1L in 1948. As the initial performance of this aircraft was poor, development continued and, in 1951, the engine was replaced with the HS 12-Z-19. However, this was still not powerful enough, and, in 1956, it was replaced once again with the Rolls-Royce Merlin 500-45. With this engine, the aircraft was renamed the HA-1112-M1L; between 1954 and 1958, 171 were built and were delivered in 1960, remaining with the EdA until 1965.

A Savoia S.79, belonging to 52 Regimiento de Bombardeo (52 Bombing Regiment), is seen in Barajas in 1942. These aircraft were initially sent to Son San Juan for the surveillance of maritime routes. This aircraft, 28-59, was the survivor of the clash with the British in the vicinity of the Baleares Islands, where another S.79 was shot down.

CASA acquired the licence to build the Ju 52 in 1942. The licensed aircraft was named C-352L and took its maiden flight in 1944; around 170 of these aircraft were built and delivered between 1941 and 1955. Additionally, CASA also acquired the licence to build the He 111H-16 in Seville. Named CASA 2.111, this aircraft took its maiden flight in 1944, and the first one was delivered to the EdA in 1945, with total production reaching about 236 aircraft. However, both aircraft had some problems during the manufacturing process, because of the lack of German engines for them; consequently, they had to re-engine them with new engines.

However, the needs of Spanish Aviation were not only fulfilled by foreign-designed aircraft. Spanish aircraft and engine manufacturers also produced original aircraft during the difficult years after World War Two. A transport aircraft named CASA C-201 Alcotán (Kestrel) performed its maiden flight in February 1949. Next, an enlarged transport aircraft named CASA C-202 Halcón (Hawk) first flew in May 1952. There were 17 C-201s (two prototypes, 12 pre-series, four series) and three C-202s manufactured (only one of the three prototypes belonged to the EdA).

Meanwhile, some old aircraft suffered the attrition of war and, consequently, in 1946 only six He 60s were in service, both of which were retired in 1953. Similarly, the last two Junkers Ju 86 Fumos, still in service during 1945, were retired in 1946.

Two lesser-known aircraft built in Spain during the later 1940s and early 1950s were the Huarte Mendicoa HM-1 (with an open cockpit; B type with a closed cockpit) and the Hispano-Suiza HS-42. At least 190 HM-1s were delivered to the EdA in 1950–55 (70 HM-1s and 120 B types) and 101 HS-42s

Left: Messerschmitt Bf 109 E-3 at San Javier AB. These aircraft were the most powerful fighters of the EdA, and they were intended to protect the border with France.

Below: After the Spanish Civil War (SCW), the Savoia S.81 was deployed by the 1ª Brigada Aérea (1st Air Brigade) of the Cantabrico Air Region. This S.81 is parked at Alcantarilla AB, Murcia.

Above left: At the end of the SCW in 1939, the EdA decided to start license-building several aircraft types, from both German and Italian design. Hispano Aviación began the manufacturing under licence of the Italian Fiat CR.32 Quarter, which was later renamed HA-132L when it was made in Spain. The military name, however, was still C.1, as it was for the CR.32s built in Italy. Seen in this image is a two-seater HA-132L deployed at the Academia General del Aire (AGA).

Above right: At the Orel AB, a Messerschmitt Bf 109 F-2 is pictured. This aircraft belonged to the second Blue Squadron that fought in the Russian skies. On the fuselage and next to the Balkenkreuz cross is the emblem of the yoke and arrows, used by the Falange Española de las JONS (Spanish Phalanx of the Councils of the National Syndicalist Offensive). (Juan Arráez Cerdá)

One of the Bf 109 Fs that Germany sold to Spain during World War Two is pictured during the 1940s. This fighter was officially designated as C.4F in the EdA but was nicknamed 'Zacuto'.

between 1946–50. Both aircraft were intended for training roles and remained in the EdA until 1958. A light aircraft named AISA I-115, also intended for training and liaison roles, was also built during this time, 200 of which were produced for the EdA.

Several Stinson 108-3 aircraft were purchased by the EdA from the US in the early 1950s and used as liaison aircraft. During this time, 108-3s and the Fieseler Fi 156 Storch were the best liaison aircraft in the EdA.

It was also in the 1950s that the EdA received its first helicopters; the Aerotécnica AC-12 and the Aerotécnica AC-13. These helicopters were manufactured in Spain, under licence, using French designs. Ultimately, only 12 and three were made, respectively.

Following the SCW, there were several seaplane types flying for the EdA. The use of these aircraft was a mixture of maritime reconnaissance and search and rescue (SAR); these tasks were mainly carried out in the waters of the Mediterranean Sea and, less frequently, around the Canary Islands.

C.4J-10 (94-28) was saved from the scrapyard by being sent to the Spanish Air Museum in Madrid. When this picture was taken, it was still in a hangar at Cuatro Vientos airfield in the late 1970s, before it became one of the stars of the museum. The EdA's Hispano HA-1112-K1Ls were painted a bluish grey and the numbers on the fuselage were written in black.

Left: After the SCW, the Grupo de Hidroaviones (Seaplanes Group) was based at Pollensa AB. It was formed by the 52 Escuadrilla, equipped with three Heinkel He 59, three Heinkel He 60s and three Arado Ar 95s (belonging to 51 Patrulla) and the 53 Escuadrilla, equipped with two Cant Z.501s and two Cant Z.506Bs. Here, a Z-501 seaplane is pictured flying over Palma de Mallorca.

Below: The Caproni Ca.310 arrived in Spain in 1938. In 1939, Spain had 16 Ca.310 aircraft intended for reconnaissance and assault. After the SCW, they were deployed to the 43 Grupo, but by 1949, they were all retired.

At the end of 1938, 10 Caproni AP.1 aircraft arrived in Spain. These were deployed to the Escuela de Caza, based at Villanubla AB. Although built as an assault aircraft, the AP.1s were used as training aircraft by the EdA and they flew until 1948, when all were retired.

The German-designed Junkers Ju 52 was renamed CASA C-352L when manufactured under licence in Spain. The aircraft successfully completed its maiden flight in 1944. Although eventually obsolete, this aircraft was the backbone of the Spanish transport branch for many years. This picture was taken in Spanish Sahara, as it was then known, on the sandy airstrip where the C-352L had to land and take off.

A C-352L rests in front of the hangar at San Javier, whilst waiting for the next mission, in 1967. Around 170 aircraft were built and delivered to the EdA between 1941 and 1955. Note the large Spanish cockades and St Andrew's Cross. This livery was common for the C-352L while in the EdA.

A Hispano Aviación HA-1112-M1L, known in the EdA as C.4K, at El Prat Airport, Barcelona. Note the two 20mm Hispano Suiza cannons, the voluminous Rolls-Royce Merlin engine and the four bladed propeller.

A beautiful photograph of C.4K 'MAPI' of the 71 Escuadrón Táctico de Cazabombarderos (71 Fighter-Bomber Tactical Squadron), taken during a visit to Palma de Mallorca in May 1964. Its official designation of C.4K-9, with UN 71-5, can be seen. 'MAPI' was one of the C.4Ks that took part in the first detachment to Spanish territories in Western Africa and the Canary Islands on 30 January 1958.

A Dornier Do 24T-3 seaplane is pictured flying from Porto Pí. The 12 Do 24s purchased from Germany increased the Spanish capabilities in rescue missions. This aircraft is painted in high visibility colours for SAR missions.

On 1 January 1943, one Focke Wulf Fw 200 C-4/U3 (F8+AS) belonging to 8./KG 40, tried to return to its base, but is pictured having to land in Sevilla. The aircraft stayed in Spain, later entering service in the EdA. No other Fw 200 flew for the EdA.

Among the types of aircraft sent to the war in Sidi Ifni were the HA-1112-M1L, the Ju 52 (CASA 352), the CASA 2.111 and the North American T-6 'Texan' two-seater trainers, although by this time the EdA already had the North American F-86 Sabre. In this photograph, we can see a Spanish-made CASA 2.111 from Ala 46.

A CASA C-1131E (E.3B) is pictured flying over Spain. The license-building of several aircraft, such as the Bücker Bü 131, known as C-1131 when manufactured in Spain, was very necessary for the training of new EdA pilots during the 1940s, 1950s, 1960s and mid-1970s. The last Spanish C-1131E was built in 1963 and, after a long life of service, they were retired in 1976.

Above: A flight of original Gotha Go 145, licence-built as CASA C-1145L. The CASA 1145 was one of the first aircraft to be licence-built following the end of the SWC in 1939.

Left: A line-up of several CASA C-1131E aircraft, based at San Javier AB. The light grey livery typical of this aircraft type can be seen here. The UN belongs to 791 Escuadrón, intended for the elementary flying school within the AGA.

After the end of the SCW, the 62 old and weary Savoia S.79s became the backbone of the bomber branch. These aircraft were very modern when they arrived in Spain at the end of the 1930s, however, they became obsolete in the following decade. This aircraft is pictured parked at La Rabasa airstrip, Alicante, in 1950.

From Spanish Civil War to 1953: The Reorganisation

Above: In 1940, only 15 Savoia S.81 aircraft were operational within the bomber branch. These aircraft, like their 'sister' the S.79, were surpassed by more modern bombers during the 1940s and were consequently resigned to more secondary roles. This aircraft was parked in Alcantarilla AB, Murcia.

Right: The S.79 not only flew as a bomber, but also as VIP transport, like the one in the picture. These aircraft fulfilled their missions during a difficult age for the EdA, as it struggled on with the old and weary aircraft it had.

Pictured here is one of five Ro.43s, and one Ro.44, that were interned and then acquired by Spain from Italy after their pilots looked for refuge in Spain after the Italian armistice of 1943. These aircraft were not put into service before the end of World War Two. They were flown for a mixture of maritime reconnaissance and SAR missions. The aircraft in this picture was based at Pollensa.

Left: Pictured here is one of the myriad of aircraft types that were used for training within the EdA, the Ro.37bis. The EdA had ten Ro.37bis aircraft deployed to the Escuela de Transformación, (Transformation School) based at Jerez de la Frontera, in 1940.

Below: A line-up of several Dornier Do 17s and a Tupolev SB-2. The eight Do 17s that the EdA had when the SCW ended were used as reconnaissance aircraft, while the 18 SB-2s were used as bombers. Note the different livery of both aircraft types.

A Spanish Fieseler Fi 156 Storch is pictured waiting for the next flight. Six Fi 156As (with no arms) were delivered from the Condor Legion to the EdA at the end of the SCW, and, in 1943, 20 new Fi 156s (A and C types) were purchased from Germany. This aircraft was used for cartographic works, within the HQ Squadron, as liaison, glider towing and utility aircraft until they were retired in 1962 and replaced by the Do 27 and CASA C-127.

Right: Around 18 Heinkel He 45s belonged to the EdA in 1940. These old aircraft were used for training duties. Most of them were deployed to 41 Grupo, based in Málaga, and San Javier AB.

Below: The role of the Escuadrilla Meteorológico (Meteorological Squad) was to survey the area of the Bay of Biscay during World War Two. This unit sent information to the Germans, who used it to their advantage both for their submarine fleets and for the Luftwaffe. This unit used three Heinkel He 111Js from February 1941 and three He 111H5s from December 1943. In this picture, we can see one He 111J in silver livery.

In 1940, the bomber branch of the EdA operated as many as 52 He 111s. Here, a He 111E is pictured parked at Agoncillo in 1941. This aircraft does not have the Spanish cockade and instead bears the arrows and yoke emblem used in the SCW.

A beautiful shot of an He 111E, painted in a tricoloured livery of dark green, light brown and sand. This aircraft bears the Spanish cockades in six positions and the St Andrew's Cross on the tail. This aircraft was considered old even at the start of World War Two.

A nice picture of a Klemm Kl 32, which belonged to the Grupo de Estado Mayor (Headquarters) until 1952; it was fully retired in 1953. Spain only purchased four of these aircraft during the SCW. Behind the Kl 32 is an He 70 'Rayo'.

After the SCW, Spain had 11 He 70s (R-2). These aircraft were often used as VIP transport, but they became obsolete in the 1940s. In this picture, this He 70 still bears the arrows and yoke emblem but not the Spanish cockade.

From Spanish Civil War to 1953: The Reorganisation

After the end of the SCW, the EdA had two types of attack aircraft: the obsolete Heinkel He 51 and the Polikarpov I-15. The He 51s, like the one pictured here, were deployed to the 31 Regimiento, based at Getafe airfield.

After the SCW, the EdA acquired several types of seaplanes: the He 59, He 60, Dornier Do J Wal and Ar 95 of German origin; the Cant Z.501 Gabbiano and Cant Z.506 Airone of Italian origin. Pictured here is an He 60 that was deployed to 52 Escuadrilla from 51 Grupo, based at Pollensa AB.

Pictured here is one of the He 59 seaplanes that had been deployed to 52 Escuadrilla. These old seaplanes did their best in SAR and reconnaissance missions for many years.

Left: After the SCW, the EdA acquired a number of reconnaissance aircraft. Among them were aircraft of Soviet, Italian and German origin; an example of the last being the He 45. In 1940, Spain had 18 He 45s deployed to 41 Grupo, based at Mola AB, Vitoria. Not long after their implementation, the He 45s were taken off reconnaissance missions but remained in active service in other branches of the EdA.

Below: At the end of the SCW, Spain had 14 Messerschmitt Bf 109 A and B types. These aircraft were already considered obsolete by the beginning of World War Two. Seen here is an example of a flagship aircraft used by the Condor Legion during the SCW. (Public domain)

Pictured here is one of the 36 Bf 109 Ds (C.4) delivered to Aviación Nacional (Nationalist Air Force) during the SCW. They were numbered from 6-51 to 6-86. Some of these planes remained in flight with the Grupo de Caza (Fighter Group) of Reus until they lost ground in 1954.

Right: A Bf 109 E-3 at El Prat Airport, Barcelona, in 1942. These aircraft were the most powerful fighters of the EdA in the early 1940s and were intended to protect the border with France.

Below: The same Bf 109 E-3 at El Prat Airport. These fighters, together with the Bf 109 Fs, were the spearhead of the EdA.

The Spanish pilots from the Escuadrillas Azules (Blue Squadrons) manned two aircraft types: the Bf 109 and Focke-Wulf Fw 190. Spanish officers chat with their German counterparts at the Toulouse-Colomiers AB; the Fw 190s, with which the 4 Escuadrilla Azul pilots trained, can be seen in the background.

One of the several Luftwaffe aircraft interned in Spain and later put into service in the EdA was this Junkers Ju 290. It took years to make this aircraft operational again.

Pictured here is a Condor Legion He 112. Along with the Bf 109 E, the He 112 was the best fighter aircraft owned by the EdA at that time, and all were deployed to Nador AB, Melilla.

A Fiat Br.20, which belonged to the Italian 23 Squadriglia during the SCW. Thirteen of this venerable aircraft fought with the Aviazione Legionaria (Legionary Aviation), of which ten survived the SCW and were delivered to the EdA.

A line up of Spanish-built Hispano-Suiza HS-42Ds, an improved and re-engined version of the HS-42B. About 100 HS-42Ds were built, and they flew within the EdA until the mid-1950s.

Two C.4s belonging to the 23 Regimiento de Caza, based at Reus AB, in the 1940s. It is noteworthy that one C.4 was two-bladed and the other three-bladed.

Three Bf 109 F-2s are lined up in the Escuela de Caza at Morón de la Frontera, Sevilla. The aircraft in the background is a Caproni AP.1.

Chapter 2

From 1953 to 1965: American Aid and Spanish Aircraft Industries Impulse

At the beginning of the 1950s, the EdA had about 900 aircraft, but only 600 were operational. The situation was dire. However, in this time of need, the US came to the aid of the EdA. On 24 September 1953, a co-operation agreement with the US was signed. This agreement meant that modern aircraft, primarily consisting of jets, began to arrive in Spain. In addition to the procurement of aircraft, this treaty also included an agreement for mutual defence, a financial assistance agreement and a defence agreement.

It was in April 1954 that the US began to supply modern aircraft to Spain. The first to arrive were Lockheed T-33A jets, alongside some Grumman HU-16As. On 30 June 1955, the first F-86F Sabre jet arrived in Getafe, Madrid, and marked a milestone in the history of the EdA. In total, 270 F-86Fs were received across five years because of the agreement.

In 1955–57, the US delivered around 200 F-86Fs, 120 North American T-6D/G Texans, 30 (increased to 48 by 1958) T-33As, five Grumman HU-16A Albatrosses, 13 Cessna L-19As, 15 Douglas C-47s, and several helicopters including four Sikorsky UH-19s, two Hiller OH-23s and one Bell 47G.

However, even before the arrival of these new and modern aircraft, the EdA was undergoing numerous changes that made it comparable with the air forces of other European countries. Spanish manufacturers now needed to compete with the American aircraft, and this helped kickstart a new wave of homegrown industry.

The EdA purchased 37 Agusta-Bell AB-47G2/G3s and Bell OH-13H Sioux helicopters, which were flown from 1962 until 1990. During their time with the EdA, these light aircraft were used extensively.

Helicopters

The UH-19 was the first helicopter that ever flew for the EdA and was, alongside the HU-16As, the main Spanish aircraft intended for SAR roles after 1955. Between 1954 and 1969, 17 UH-19 A, B and D types, produced by both Sikorsky and Westland Aircraft (which was known as the Westland Whirlwind), flew for the EdA in a utility role.

Fighters

The last F-86F was delivered to the EdA in July 1959; in total, Spain acquired 18 F-86F-20s, 155 F-86F-25s, 32 F-86F-30s and 65 F-86F-40s. All the F-86Fs came from surpluses of the USAFE and, although beginning as different variants, all were eventually updated in Spain to the F-86F-40-NA.

New Units

Thanks to these new aircraft, the EdA were able to establish a variety of new units. One of these new units was the Escuela de Reactores (Jet Training School), which was formed in 1954 with 60 T-33s at Talavera la Real, Badajoz. Furthermore, five Alas de Caza (Fighter Wings) were formed from 1956

The CASA C-207 was a Spanish-built short- and medium-range cargo and troop transport aircraft. While it was undeniably an important milestone for the Spanish aeronautical industry, only ten C-207A (troop transport) and ten C-207C (cargo transport) aircraft were ever built.

A Lockheed F-104G landing at Torrejón de Ardoz AB, Madrid. The F-104G arrived in Spain in 1965 and represented a great improvement in Spanish air power, as, until that year, the most modern jet had been the F-86. A pilot compared the F-86 and F-104 performance, saying that the first was like a car and the second was like a racing car. For five years, the F-104G remained the only supersonic interceptor aircraft in the EdA.

onwards. This aircraft soldiered on in this unit until 1973, when they were used as simple trainers. These Alas de Caza were stationed in the following locations:

- Ala de Caza 1, based at Manises Air Base (AB), Valencia, formed of 11 Escuadrón and 12 Escuadrón
- Ala de Caza 2, based at Valenzuela AB, Zaragoza, formed of 21 Escuadrón
- Ala de Caza 4, based at Son San Juan AB, Mallorca, formed of 41 Escuadrón
- Ala de Caza 5, based at Morón de la Frontera AB, Sevilla, formed of 51 Escuadrón
- Ala de Caza 6, based at Torrejón de Ardoz AB, Madrid, formed of 61 Escuadrón

In 1956, an aerobatic unit, equipped with four F-86s from Ala 1, was created and known as the Patrulla Ascua. These aircraft were painted in a special silver scheme, which included a red lightning bolt against a yellow background, signifying the Spanish flag.

Sixty T-6s were deployed to the Escuela de Pilotos (Pilot School; specifically to 793 Escuadrón), which was formed in August 1955 and was based in Matacán AB. Furthermore, the 22 C-47s delivered to the EdA, the last one arriving in November 1957, were deployed mainly to the Ala de Transporte 35 (Transport Wing), which was formed in 1955.

Trainers

The EdA also needed a new basic trainer aircraft to replace the old types, and, once again, turned to the US for a solution. The chosen aircraft was the Beechcraft T-34 Mentor. It was in service from 1958,

Four F-104Gs are seen flying over the Entrepeñas and Buendía swamps in April 1965. The two closest F-104Gs have the UN of 161 Escuadrón painted on the forward fuselage, however, the other two aircraft lack it. It is noteworthy that the flaps are in take-off position because the aircraft needed to be as slow as the Lockheed T-33 taking the picture.

Although the F-104 has had many accidents in its global military career, in Spain no F-104G was ever lost. Indeed, during their operational service, Spanish F-104Gs achieved 1,705,915 flight hours between 1965 and 1972 without losses.

invariably in 791 Escuadrón, and spending a short amount of time in 792 Escuadrón. In 1967, a further nine T-34s were purchased from the US and subsequently upgraded in Spain to the Spanish standard. The 26 T-34s acquired were used as elementary training aircraft in the Academia General del Aire (Air General Academy; AGA) based at San Javier AB, with success until 1987 when they were replaced; the last of this type flew until 1988.

Gliders
Several LET Kunovice L-13 Blaniks were purchased by the EdA for glider units in the mid-1960s. In the EdA, the L-13 played an important role in the training of pilots. Another one was the Scheibe SF-25 Falke.

Transport and Liason
In 1959, six American-built Douglas C-54 Skymasters arrived in Spain, with five deployed to Ala 35, based at Getafe AB, and one to the Grupo de Estado Mayor (Staff Group).

Spanish Industry
Thanks to the increase in modern aircraft that accompanied the agreements with the US, some Spanish-built types began to appear obsolete. One such aircraft was the HA-1112 M1L, which by this time seemed far too outdated to perform its role as a fighter within the EdA. The EdA purchased the Spanish-built AISA I.11B Vespa liaison aircraft, also known as the AISA I.115, when designed for a

A lonely F-104G is pictured taking off from Torrejón de Ardoz AB, showing the light grey livery on all but the white wings. At the time, this aircraft belonged to 161 Escuadrón, although, from 29 November 1967, 161 Escuadrón became 104 Escuadrón.

training role. CASA manufactured a new and improved transport aircraft named CASA C-207 Azor (Goshawk) and Hispano Aviación developed a basic trainer named HA 100 Triana, although it never was mass manufactured. With the arrival of the T-6s, only two HA 100s ever flew and an order for 40 aircraft was cancelled early on.

Another Spanish-built aircraft was the Dornier Do 27. The Do 27 was a superb liaison aircraft from German design and licence-built in Sevilla. The first of the 50 manufactured by CASA was delivered in December 1959. Its name in the EdA was initially L.9 (L for liaison), although, on 19 April 1978, the name was changed to U.9, after it took on a utility role. As a multi-purpose aircraft, it worked in many different units including 803 Escuadrilla in an SAR role, 432 Escuadrilla de Enlace (Liaison Squadron), 407 Escuadrilla de Enlace, glider towing and parachuting units, and more. With the Do 27, the venerable Fi 156s and 108-3s were replaced.

During 1964, the EdA created the Tactic and Transport Aviation division. The Ala de Transporte 35 (Ala 35) was made up of 351 Escuadrón and 352 Escuadrón; the C-207 was distributed to the former, and the C-54 to the latter.

The best Spanish manufactured aircraft in this time frame was the Hispano HA-200 'Saeta' (Arrow). It was the first jet manufactured in Spain and was designed by Willy Messerschmitt. It took its maiden flight on 12 August 1955 and first flew for the EdA in June 1960, subsequently being deployed to different units in the early 1960s. The new aircraft was intended to have the jet trainer role, but, primarily because of the good performance, an attack version was manufactured soon after. Within the EdA, the aircraft was named C.10 at the beginning, but, after 1978 and in its new role, was known as A.10. The maiden flight of the HA-200 was on 16 August 1955. An improved version named HA-220 'Super Saeta'– known as C.10C at the beginning and A.10C since 1978 – was intended for tactical and close support role and was pressed into service at the beginning of the 1970s. From the end of the 1950s until 1980, 97 HA-200As and Ds and 25 HA-220Es flew in the EdA.

In 1965, as part of the Mutual Defense Assistance Program, Spain received 18 Lockheed F-104G Starfighters and three TF-104Gs (one of which was bought by the EdA). These fast aircraft were deployed to 104 Escuadrón, and then to 161 Escuadrón, both of which belonged to Ala 16 and were based at Torrejón de Ardoz. The pilots for the F-104Gs were recruited from the F-86 pilots, with at least 500 flying hours in this aircraft. While flying for the EdA, this aircraft achieved a record, as, during seven years in service, no aircraft was lost, so, in 1972, the F-104s were gifted to Greece and Turkey at the suggestion of the US.

While the EdA was in possession of a handful of helicopters, it strove to have a greater number in its arsenal. Consequently, 12 Aerotécnica AC-12 helicopters flew within the EdA in 1961–63, and six Aerotécnica AC-14s flew during 1963. An important acquisition was the 37 Augusta-Bell AB-47G2/G3 and Bell OH-13H Sioux helicopters, which were flown by the EdA between 1962 and 1990. Additionally, three Augusta-Bell AB-47J-3B-1 helicopters flew with the EdA from 1962.

A T-33A from 41 Grupo. This aircraft type flew in this Grupo until it was replaced by the CASA C-101 in 1981. The aircraft has an all-silver livery, as was usual for the T-33s; note the Ala 2 emblem of a tiger. Today, this emblem is used by Ala 15 with the motto 'Quien ose paga' ('Who dares pay').

The first T-33A arrived in Talavera la Real in March 1954. This was the first jet in the Escuela de Reactores (Jet School) and first flew from there on 24 March 1954. The pictured aircraft belonged to 732 Escuadrón, based at Talavera la Real AB.

The first F-86 arrived at the Escuela de Reactores in October 1958. Once there, it formed part of the the Escuadrón de Aplicación y Tiro (Shot and Application Squadron), which provided training for new pilots. This aircraft belonged to 732 Escuadrón.

When the F-86 arrived in Spain, it was necessary to train the new pilots. Before the F-86 arrived, the best fighter the EdA had was the Bf 109 F.

Spanish F-86s were painted in an all-silver livery, with Spanish cockades in six positions and the St Andrew's Cross on the tail. The colour of the band painted on the nose indicates the unit.

During its service, the Spanish F-86 accumulated more than 360,000 flight hours. For 17 years, the F-86 was the masterpiece of the Spanish air defence. When the last of this aircraft was retired in December 1972, it was replaced with newer aircraft such as the Dassault Mirage III, McDonnell F-4, F-104 and Northrop F-5 Freedom Fighter.

A T-33A from 101 Escuadrón, based at Manises AB, Valencia. The T-33As were used as trainers in a variety of missions except for tactical or air defence, the latter of which was fulfilled by F-86s.

In this picture, three of the first jet aircraft types that flew for the EdA can be seen. In the foreground is an F-86 from 102 Escuadrón (previously 161 Escuadrón), and behind it is a T-33A from 41 Grupo and a two-seater F-5 (SF-5B).

A line-up of F-86s from 111 Escuadrón, awaiting the next mission. Despite the weary status of these aircraft when they arrived in Spain, they were still an important step forward for the EdA.

In 1956, an aerobatic team with jet aircraft was formed, later known as the Patrulla Ascua. At the beginning, it flew four F-86s from Ala 1, although more aircraft were used later. These aircraft were painted in a special silver scheme, which included a red lightning bolt against a yellow background, signifying the Spanish flag. In this picture, we can see a six-ship formation performing for a show.

The first Beechcraft T-34 Mentor arrived in Spain in November 1957. The 26 T-34s purchased were used as elementary training aircraft in the AGA with success until 1987, when they were replaced.

The last T-34 formation. This aircraft type was a masterpiece for the training of new pilots for the EdA. As well as training, this aircraft replaced the venerable HS-42 in navigation, shooting and photography missions.

From 1953 to 1965: American Aid and Spanish Aircraft Industries Impulse

Right: A superb picture of a Do 24T-3 flying close to a Do 27 during a SAR mission. Both Dornier aircraft flew within the EdA with great success for many years. This shot was taken in 1966.

Below: The Do 27 was a superb liaison aircraft, licence-built in Sevilla. Initially, the main role of this aircraft was liaison, and it was consequently named L.6. However, after some years, it changed to a utility role and was renamed U.9. The picture was taken at San Javier AB – note the C-101 in the background.

This Do 27 was intended for SAR missions and is painted with high visibility markings. These aircraft were deployed to the 803 Escuadrilla, where it worked with the Do 24T-3 seaplane. The shot was taken at Son San Juan AB.

UH-19s, built by both Westland Aircraft and Sikorsky, became the first helicopters to fly within the EdA. They flew for the EdA between 1954 and 1969 in utility and SAR roles. The helicopter in the picture was built by Westland Aircraft and has distinctive SAR markings.

Left: A beautiful shot of the EdA's first anti-submarine warfare (ASW) aircraft. The upper surface of this Grumman HU-16 Albatross is grey, while its under surface is white – this two-tone livery was specifically designed for ASW missions. The aircraft belonged to 611 Escuadrón, as can be seen from the emblem of a winged thunderbolt over a yellow submarine and a crossed chain.

Below: In May 1954, seven HU-16 A series aircraft arrived in Spain and were deployed to 55 Escuadrilla and 56 Escuadrilla for SAR roles. In 1971, the EdA purchased six HU-16 B series aircraft, with a better performance rating than the A series. In this picture, taken in 1966, there are three HU-16s lined up at Son San Juan AB.

From 1953 to 1965: American Aid and Spanish Aircraft Industries Impulse

Right: One glider aircraft used by the EdA was the Scheibe SF-25 Falke. Thanks to these aircraft, pilots could start their training before handling other engine aircraft. This aircraft from the AGA was captured by the photographer in 1992.

Below: The EdA had 67 Douglas DC-3s from different series, which were primarily deployed to Ala 35, Ala 37, Ala 46, Grupo de Estado Mayor, Escuela de Polimotores (Multi-engine Aircraft School) and Escuela de Paracaidistas (Parachute School). The EdA purchased its first two Douglas C-47s in 1947 and the last one flew until 1978. This picture was taken at Villa Cisneros (now Morocco, but then a Spanish territory).

Most of the DC-3s arrived in Spain after the 1956 agreement with the US, and they were primarily deployed to Ala 35, where 22 of these aircraft were used for transport. Another batch of DC-3s were delivered in 1963–65 and were deployed to the newly established Ala 37, where 30 aircraft also took on a transport role. The DC-3 in the picture was deployed to the Escuela Superior de Vuelo (Upper School of Flight, part of the AGA).

Above: One of the several LET Kunovice L.13 Blanik gliders purchased by the EdA from the mid-1960s. The aircraft was deployed to the AGA, and one is seen here, close to a C-101, at San Javier AB.

Left: The T-34s were useful training aircraft and were eventually replaced by the T-35s in 1987.

A Douglas C-54, belonging to Ala 35, is seen waiting for its next mission. The EdA purchased 17 C-54s, which were deployed to Ala 35 and within 91 Grupo for VIP transport. The service life of these aircraft was from 1959 until 1978, when the last C-54 was retired.

From 1953 to 1965: American Aid and Spanish Aircraft Industries Impulse

Above: A Spanish-built Hispano HA-220 'Super Saeta' is pictured close to the Jerez AB, Andalusia. This aircraft was a one-seat improvement of the Hispano HA-200 'Saeta', intended for close attack missions. This aircraft entered service in 1973 in 406 Escuadrón and 214 Escuadrón and is camouflaged in sand and green with light blue under surfaces.

Right: On 30 June 1955, the first F-86F jet arrived at Getafe, Madrid, marking a milestone in the history of Spanish aviation. This jet is painted in Patrulla Ascua (Eagle Patrol) livery. (Garrapata)

The EdA purchased 60 North American T-6 Texans in 1954, which served in the Escuela Básica (Basic School), based at Matacán AB. Due to the success of the T-6, a batch of 60 T-6Gs was purchased at the end of the 1950s. Later, during the Sidi Ifni War, Spain could not use American jets, so they had to buy a new batch of 81 T-6s from France. The aircraft in this picture has a livery of sand, dark earth and black-green, which was used on Spanish aircraft in African territories.

Left: When it was being used as a training aircraft, the standard livery of the T-6 was high visibility yellow. In 1962, the C.6s were sent to the AGA from the Escuela de Vuelo Básico, where they flew until 1982 when they were replaced by the C-101. A formation of E.16s from 793 Escuadrón pose for the photographer.

Below: The first jet ever built in Spain was the HA-200 'Saeta'. This aircraft was designed by the well-known Willy Messerschmitt in 1955. This two-engine aircraft was built at Triana quarter in Sevilla and performed superbly. In total, 117 'Saetas' flew for the EdA: two prototypes, five pre-series, 30 HA-200As, 55 HA-200Ds and 25 HA-220 'Super Saetas'.

Whilst in the EdA, the attack version of the HA-200 'Saeta' was named A.10, the fighter type was named C.10, although in 1978 this changed to A.10, and training types were named E.14B. At the beginning, these aircraft were based at Villanubla, Matacán, San Javier, Gando and Morón de la Frontera (where they joined the last HA-200 in service with 204 Escuadrón until 1981 when they were retired). This HA-200 is photographed flying over Canary Islands. Note the typical 'Saeta' livery: anti-corrosion grey, with red around the intakes and a bolt outlined in white.

Chapter 3

From 1965 to 1985: Improving the Breed

Unit Restructuring

The EdA had now started a new age, and it was intent on both optimising its new, modern aircraft fleet and acquiring replacements for its older fleet. As part of this, in 1965, the EdA was reorganised and the Alas de Caza changed their names:

- Ala de Caza 1 became Ala 11, and it consisted of 101 and 102 Escuadrones
- Ala de Caza 2 became Ala 12, and it consisted of 121 Escuadrón
- Ala de Caza 5 became Ala 15, and it consisted of 157 Escuadrón
- Ala de Caza 6 became Ala 16, and it consisted of 102 Escuadrón

In 1967, further changes within the EdA were required, largely owing to the remnant damage done to the fleet through wartime attrition. The most important change was the disbandment of the Alas that were replaced by Escuadrones. For example, Ala 11 became Escuadrón 101, Ala 12 to Escuadrón 102, Ala 15 to Escuadrón 103 and Ala 16 to Escuadrón 104.

Attack and Fighter

Further positive changes would arrive soon for the EdA. Spanish authorities agreed with Northrop Grumman to licence-build the F-5 Freedom Fighter. CASA was the aircraft manufacturer selected to build 70 F-5s; they were to produce 36 single-seater SF-5As and 34 two-seater SF-5Bs. The first Spanish-built F-5 took its maiden flight in May 1968. Following this, 18 of the SF-5As were modified to be reconnaissance aircraft named SRF-5A. As new modern fighters arrived in Spain, the role of the F-5 was changed to an attack aircraft. Even today, a small group of modernised and upgraded F-5s continue to fly for the EdA.

The E.26 Tamiz was a Chilean aircraft purchased by the EdA, in the same way that the Spanish C-101 was sold to the Chilean Air Force. This aircraft was a modified version of the Piper 'Aztec', which, by November 2017, had flown 100,000 flight hours in its trainer role in Spain. In 1987, 41 T-35s were deployed to the Escuela Elemental (Elementary School) from the AGA. These aircraft were the first step for cadets flying the AGA's trainer jet, C-101.

The first SF-5 equipped unit was 202 Escuadrón, based at Morón, in January 1970. One year later, the unit was deployed to the Escuela de Reactores and renamed 731 Escuadrón. The same thing happened to the second unit equipped SF-5s; 204 Escuadrón was turned into 732 Escuadrón, which also belonged to the Escuela de Reactores. At the end of 1971, the last Spanish SF-5 was delivered to the EdA. In Ala 21, the first SF-5s were operated together with the HA-220.

The EdA was also looking to replace its F-86s and had its eye on two aircraft: Dassault Mirage IIIs and McDonnell Douglas F-4 Phantoms.

In 1970, Spain acquired 24 Mirage III EE single seat fighters and seven Mirage III DE two-seaters. These French-designed aircraft were deployed to the Ala de Caza 10 (comprising Escuadrón 101 and Escuadrón 103) based at Manises, Valencia. Shortly after, in May 1971, this Ala was renamed as Ala 11 and now held 111 Escuadrón and 112 Escuadrón. The Mirage III flew for the EdA for 21 years. When the EdA purchased the Mirage IIIs from France, they also sent a message to the US that Spain no longer depended only on American help.

Above: During training, it was mandatory for Spanish pilots to fly an E.26. The word 'Tamiz' means 'sieve' and it was so named as it was an aircraft intended to differentiate the skill of the pilots. The aircraft in the background is a Spanish-designed CASA C-212 Aviocar.

Left: E.26s flying in tight formation over San Javier AB, where the AGA is based.

The other aircraft Spain acquired was the F-4, which was designed to replace the battered F-86Fs and F-104s. The EdA was interested in the F-4E, but budget constraints forced it to accept the F-4C. Thus, in the end, Spain acquired 36 F-4Cs, alongside three Boeing KC-97Ls – a strategic tanker aircraft that Spain now needed – and two C-97 transport aircraft, which were only used as a replacement source for the KC-97Ls. The F-4Cs were not brand new, indeed several were Vietnam veterans and others came from the 81st TFW of USAFE. The first Spanish F-4 arrived in Spain on 19 February 1971.

When the F-4Cs were in Spain, they were deployed to 121 Escuadrón 'Poker' and 122 Escuadrón from Ala 12. Of course, the KC-97Ls and C-97Gs were incorporated into this Ala too.

After several accidents, in 1978, the EdA acquired four F-4Cs to replace those that had been written off. Sadly, as all aircraft do, the F-4 eventually became outdated, and they were retired from service in 1991. However, the story of this superb aircraft had much to offer in the interim.

At last Spain had an Air Force able to defend the Spanish skies like never before. However, the EdA still wished to add a new fighter. Again, Spain looked to France to buy the new Dassault Mirage F-1, and 15 F-1Cs were bought in 1972. These aircraft were deployed to 141 Escuadrón 'Chico' from Ala 14, based in Los llanos. During 1976, nine F-1s were delivered to Spain, followed by 48 F-1s in 1978. This aircraft was designed to fulfil the Spanish requirements for air defence, and such was the confidence in this fighter that new F-1s, from different batches, were delivered until March 1995: F-1BE, F-1B, F-1CE, F-1C, F-1DDA, F-1EDA and F-1EE. Regardless of the variant and the improvements done to them, the Spanish F-1s were always called C.14 when they were a single-seater and CE.14 when a two-seater. In total, 91 F-1s flew with the Spanish units, and, although they were retired from service, they are still remembered nostalgically because of their valuable work in Spain.

With the greater number of aircraft available, a new Escuadrón was created: 142 Escuadrón. F-1s were flown by 141 Escuadrón and 142 Escuadrón from 1975 and 1980, respectively. The F-1s were also deployed to 462 Escuadrón in 1982, which was based in Gando. The aircraft was also temporarily deployed to Ala 11.

An E.26 formation from the AGA, displaying nose, tails and wing tips painted in red, as was usual for trainer aircraft. The cadets of the AGA have their first experience in the skies with this aircraft during their third year of training.

Seaplanes

In 1971, Spain purchased two Canadair CL-215 amphibious aircraft, specially developed as water bombers. During the next several years, the total number of Canadairs in service increased to 14, although belonging to the improved CL-215T type. These aircraft were deployed to 43 Grupo, based at Torrejón de Ardoz and Pollensa.

The next version of the CL-215 was the Bombardier 415 and, as of 2006, Spain had acquired four, all of which were also based in Torrejón de Ardoz. Some of the UD.13s and 415s do not belong directly to the EdA but are operated by it.

Another area in which Spain required more aircraft was anti-submarine warfare (ASW). Thanks to the Mutual Defense Assistance Program, the EdA acquired some HU-16As in 1955–57. The rough aircraft did its best, however, the EdA soon began to demand new aircraft from the US. Therefore, from 1954 until 1971, it received 26 HU-16s from several origins and batches. These seaplanes were intended for SAR and ASW missions and were based mainly at Son San Juan AB.

Left: **Eight Sud Aviation SA 319B Alouette IIIs flew for the EdA between 1973 and 1982. This helicopter has SAR missions as its focus, but it was sometimes also used as a VIP transport aircraft.**

Below: **The SA 319Bs were deployed to 803 Escuadrón. This escuadrón, together with 402 Escuadrón, constituted the two units of the Ala 48, which was formed on 19 February 1992.**

Above: The Piper PA-23s purchased by the EdA successfully fulfilled their role for many years. This Piper is parked at Getafe, and all of those bought were of the Turbo-Aztec type.

Right: Two twin-engine Beechcraft King Air C90 aircraft entered service in 1974, intended for a training role. However, this aircraft continues to fly for the EdA with 409 Escuadrón (CECAF) as a VIP transport and liaison aircraft. The aircraft in the picture is landing at San Javier AB.

With the operational life of its HU-16s coming to an end in the early 1970s, the EdA looked for a natural substitute for the aircraft, and the Lockheed P-3 Orion was chosen. On 25 July 1973, Spain received three P-3As from the US, which were intended for anti-submarine combat and maritime patrols. Soon after, the EdA showed its interest in more of these aircraft and, in 1979, it decided to lease four P-3As from the US Navy, which were returned in 1989, after the purchase of five P-3Bs from Norway.

In 1992, P-3s were deployed from Ala 22, based at Jerez de la Frontera AB, to Ala 21 based at Morón AB. Two P-3As and five P-3Bs were flown for ASW and maritime patrol (intelligence missions, ship identification, etc), fishing protection, operations against drug trafficking and illegal immigration, and anti-surface combat (detection, location, monitoring and attacks against enemy ships and submarines) roles. Nowadays, the three active aircraft are deployed to Ala 11, based at Morón AB.

In 1953, Servicio Aéreo de Rescate (Air Rescue Service) was officially established. It was initially equipped with German-made Do 24T-3s, however, new aircraft soon arrived. The Servicio Aéreo de Rescate included 801 Escuadrón, with an inventory of HU-16Bs, CASA C-127s and Agusta-Bell AB205s; 802 Escuadrón, with AB205s and HU-16As; and 803 Escuadrón, with C-127s, Bell 47 Js, AB206As and Aérospatiale SA-319B Alouette III helicopters. One Fokker F27, intended

Above: A Beechcraft Baron B55 is pictured landing at San Javier AB. In 1972, the EdA purchased seven B55s intended for instrumental flight training. These were first deployed to 744 Escuadrón, based at Matacán AB, and then, in 1975, were redeployed to 42 Grupo. These aircraft were retired in 2004.

Left: The EdA purchased 30 Beechcraft 33C Bonanzas, which it named E.24A, intended for liaison duties (although they were sometimes used as basic trainers). Later, it added 16 improved aircraft named E.24B, which still fly in training and liaison roles. The 33C in the picture is landing at San Javier AB.

for SAR missions, was acquired in 1979. This aircraft was deployed to 802 Escuadrón and fulfilled its duties until it was replaced in December 2013 by the Spanish-built CASA CN-235 VIGMA, designed for maritime surveillance, and two Aérospatiale SA 330 Puma helicopters.

Helicopters

In 1960, the Escuela de Helicópteros (Helicopter School), based at Cuatro Vientos, and including Escuadrón 751 and Escuadrón 752, was equipped with Spanish-built AC-12s, Bell 47Gs, UH-19As, Hughes 269Cs and Bell UH-1Hs.

The new acquisitions included Agusta-Bell AB205s, Agusta-Bell AB206s, Aérospatiale SA-319B Alouette IIIs, H-269Cs and SA 330s. Between 1965 and 1993, 54 AB205s and UH-1H helicopters flew within in the EdA. Between 1970 and 1982, four AB206 helicopters flew within the EdA. Eight SA-319B helicopters flew for the EdA between 1973 until 1982. Taking on a training role, 17 H-269C helicopters flew within the EdA between 1978 and 2001. In 1973, the EdA purchased nine SA 330 helicopters, which were deployed to 801 Escuadrón in an SAR role.

In 1982, the EdA purchased 12 Eurocopter AS332 Super Pumas, which arrived in Spain on 22 December 1982. Ten were SAR helicopters and two were VIP transport aircraft. Currently, 14 AS332s are in service in the EdA, eight AS332 B1s and six AS332 C1s. Five of these aircraft are deployed to 802 Escuadrón on SAR missions; another five are deployed to the 803 Escuadrón from Ala 48, based at Cuatro Vientos. The other four AS332s are VIP transport helicopters and are deployed to 402 Escuadrón from Ala 48. This unit also has two Eurocopter AS532 Cougars intended for VIP transport, which were acquired in 2004.

Transport

Spanish aircraft manufacturers continued working in order to improve the EdA fleet from within. In March 1971, a two-engined transport aircraft designed by CASA, C-212 Aviocar, took its maiden flight. This aircraft was intended to fulfil several roles, primarily as light transport for goods, paratrooper platforms, troops or passengers and as a replacement for the obsolete C-47, CASA-352L and/or CASA C-207. This small aircraft was a success and many of them were sold to different countries. The EdA acquired at least 93 C-212s in different versions across several years, with the first delivery in 1974. Today, the C-212 is operated in Ala 37, CLAEX (Centro Logistico de Armamento y Experimentación/Logistics Centre for Armament and Experimentation) and 47 Grupo Mixto.

The EdA was still looking to replace its venerable C-47s, which were beginning to show their age after so many years flying for Spain. The chosen replacement was the de Havilland Canada DHC-4 Caribou. In 1967, the EdA acquired six brand-new DHC-4s, which arrived in Spain on 24 December 1967, and which were deployed to 372 Escuadrón, while 371 Escuadrón kept its C-47s. The aircraft successfully improved the transport capabilities of the EdA and, in 1970, a further six DHC-4s were purchased, followed by another 18 in 1981; notably, the latter group came from the Maryland Air National Guard, and could be characterised by the meteorological radar in the aircraft's nose. At least 30 DHC-4s were purchased by the EdA between 1967 and 1981.

By 1974, the DHC-4s had made their home in the Villanubla AB, where Ala 37 was stationed. When this aircraft too became obsolete, it had to be replaced within the Ala 37 inventory; the substitute was the Spanish-built CASA CN-235, which arrived in July 1991.

In 1979, the EdA purchased one Fokker F27 Friendship, intended for SAR missions. This aircraft was deployed to 802 Escuadrón and was retired in December 2013 after more than 50,000 flight hours and 34 years of service. Note the Spanish flag painted on the tail.

An interesting picture of the only Dornier Do 28A-1 purchased by the EdA in 1965. At the beginning, it was used as a liaison aircraft and named L.14, but later it was classified as utility aircraft and renamed U.14. Fortunately, this aircraft is preserved in the Museo de Aeronáutica y Astronáutica (Spanish Air Museum) at Cuatro Vientos, Madrid, although the initial light grey livery has been changed to metallic.

The real hero of the transport branch of the EdA was the Lockheed C-130 Hercules. This superb aircraft had long been desired by the EdA, and on 18 December 1973, the first of four C-130Hs landed at Zaragoza AB. Subsequent purchases were as follows: in 1976, three KC-130H tankers arrived; in 1979, two C-130Hs; in 1980, 3 C-130Hs; and in 1988, one C-130H-30. The C-130 became the backbone of 301 Escuadrón de Transporte, based at Valenzuela AB. The 301 was formed in 1974, and when the KC-130H arrived, the EdA was able to disband 123 Escuadrón, which was still operating obsolete KC-97s.

In 1978, 301 Escuadrón was absorbed into Ala 31, which already held 311 Escuadrón and 312 Escuadrón. The C-130 was officially withdrawn from active service on 31 December 2020.

Training

The training aircraft used in units such the AGA, the Escuela de Polimotores, based in Matacán AB, or the Escuela de Reactores, based at Talavera la Real AB, cannot be overlooked. The AGA included 791 Escuadrón, which was dedicated to basic training using T-34As and Beechcraft Bonanza F33Cs; 792 Escuadrón, which was dedicated to basic training using C-212Bs, I-115s and C-47s; and 793 Escuadrón, which was dedicated to conversion training using HA-200As and T-6Gs.

The F33C was ordered by the EdA in 1974, in two batches of 12 and 18 aircraft, respectively. The EdA named this light aircraft the E.24 and it was deployed to different units, including 791 Escuadrón (the Escuela Aeronautica/Aeronautical School), based in Matacán AB, and 421 Escuadrón from Ala 42. Some of the venerable F33Cs were retired, however, 16 improved aircraft still fly for 42 Grupo, based at Villanubla AB, in training and liaison roles.

In 1975, the EdA looked for a new jet trainer to replace the T-6s, HA-200s and T-33s, and agreed a contract with CASA to develop a new aircraft named CASA C-101 Aviojet. Initially, the EdA acquired 72 aircraft, but this number later increased to 88. The first C-101s arrived at AGA on 17 January 1980 and were deployed to 793 Escuadrón. On 23 October 1981, the C-101 was deployed to 41 Grupo (which comprised 411 Escuadrón and 412 Escuadrón), based at Zaragoza AB, to replace the T-33s. Five years later, on 15 July 1986, the C-101s were deployed to Ala 74, based at Matacán AB. One of the most interesting units where the C-101 flew, and continues to fly, is the Patrulla Águila, the Spanish aerobatic team that follows the path started by the Patrulla Ascua. This unit was formed on 4 June 1985 with teachers from AGA and, happily, it still operates today.

This aircraft was intended to fulfil a basic/advanced training role and had a limited attack capacity within the EdA during the 1980s and 1990s, although the E.25 continues to fly for Spain in the AGA, CLAEX and Patrulla Águila.

The other trainer aircraft, intended only as a basic trainer, was the ENAER T-35 Pillán, renamed in Spain as E.26 Tamiz. This aircraft was acquired to be built in Spain from kits supplied by the Chilean manufacturer, ENAER. In 1985, E.26s were manufactured, which fully replaced the old Beechcraft T-34 'Mentor' from 1987 and were deployed to the Escuela Elemental, where they remain today. However, this aircraft will be replaced shortly by the Pilatus PC-21s, as the EdA has purchased 24 to be delivered in the next two years.

Additional aircraft

The Escuela de Polimotores, which included 744 Escuadrón and 745 Escuadrón, was equipped with Beechcraft Baron B55s, King Air A100s, F33As, Beechcraft C90s and C-47s. The Escuela de Reactores included 731 Escuadrón and 732 Escuadrón, which, after the T-33 was withdrawn from service, was equipped with SF-5Bs.

Only one Dornier Do 28A-1 was ever acquired by the EdA, and it was purchased in 1965. Initially, it was the personal aircraft of the Governor of Equatorial Guinea, then later transferred to the EdA, where it was deployed to 604 Escuadrón, although it changed both unit and name during the following several years.

The EdA purchased five Dassault Falcon 20s, operating first in 1970. The aircraft were intended mainly as VIP transport deployed to 45 Grupo, based at Torrejón de Ardoz AB. They flew within this capacity until 2015.

Right: PA-23s were operated by the EdA for training and liaison purposes and were deployed to 902 Escuadrón and 905 Escuadrilla. In 1972, six PA-23s were purchased and named E.19. This shot of E.19-6, belonging to 912 Escuadrón, was taken at Getafe airfield.

Below: This P-3 is parked at Son San Juan. There are seven P-3s currently undergoing an improvement programme to extend their operational life and enhance performance. The three P-3s in active status all came to the EdA as P-3Bs from Norway, however, they have been improved to P-3M standard.

The Lockheed P-3 Orion is a very important aircraft for the Spanish sea and air control. This aircraft still flies for the EdA in ASW and maritime patrol, fishing protection and anti-surface warfare roles.

Left: In 1967, the EdA acquired six brand new de Havilland Canada DHC-4 Caribou, which arrived in Spain on 24 December of that year, and were deployed to 372 Escuadrón. These aircraft developed an important role within the transport branch during their 24 years of service.

Below: A DHC-4 is parked in Melilla Airport, awaiting its next mission. Note that, on this aircraft, the St Andrew's Cross is very small and is not in its usual place at the upper rear of the tail.

From 1965 to 1985: Improving the Breed

In 1981, 18 DHC-4s were purchased from a surplus of the Maryland Air National Guard. This batch was characterised by the meteorological radar in the aircraft's nose, as seen here.

Pictured here is one of 18 Northrop SF-5A Freedom Fighters that was modified to be a reconnaissance and attack hybrid; for this combination of roles, the EdA named this type of aircraft AR.9. Upon delivery to the EdA, these aircraft had a silver livery, which, after several years, was replaced by a tricoloured one.

As new modern fighters arrived in Spain, the role of the F-5 was changed to attack aircraft and its given name of C.9 was changed to A.9. The aircraft in this picture is painted in a beautiful tricoloured livery most suitable for that role.

A Mirage III from 111 Escuadrón is pictured flying near Manises AB, where this aircraft type was based. From May 1971, they were deployed to Ala 11, comprising 111 Escuadrón and 112 Escuadrón. In this picture, note the typical green and grey Mirage III livery and the large St Andrew's Cross.

Left: Some Mirage IIIs pictured flying over Spain. The acquisition of these French-designed aircraft represented a change in how and where the EdA sourced its combat aircraft, which had, until then, come exclusively from the US. On 30 June 1992, the Mirage III was retired from the EdA.

Below: One of the seven Mirage III DE two-seaters purchased by the EdA. This aircraft, which is pictured flying close to Manises AB, has a small St Andrew's Cross painted on the tail. When the most modern Dassault Mirage F-1 arrived in Spain, this powerful fighter was replaced.

From 1965 to 1985: Improving the Breed

Undoubtedly, the arrival of the McDonnell Douglas F-4C Phantom in Spain in 1971 was a very important improvement in Spanish air power. Thanks to this new fighter, the F-104G could be retired in 1972. In the picture, this F-4C, based at Torrejón de Ardoz AB, waits for the next flight.

When flying with the EdA, the F-4C always had the same livery: dark green, green and light brown in a lizard pattern. Note the Ala 12 emblem of a black cat head with the motto 'No le busques tres pies' ('Don't look for the three feet').

A Spanish F-4C is pictured flying close to two US F-16s. We can see from the 'TJ' on the tail that the aircraft are based at Torrejón de Ardoz AB. Although these aircraft were already classed as 'old warriors' when they arrived in Spain, they performed admirably during the years that they flew for the EdA. Although all the F-4Cs were ostensibly retired in 1989, the last of this type, named C.12-37, was officially retired in March 1991. Spain was the last country to retire its old F-4Cs.

Two F-4Cs are ready for a scramble rest at Torrejón de Ardoz AB, with the full moon watching them. All F-4Cs were deployed to 121 Escuadrón 'Poker' and 122 Escuadrón from Ala 12.

This history of the F-4Cs in Spain did not finish with the retirement of this type, because, in 1989, 18 RF-4Cs, named CR.12s, were purchased. These aircraft were all painted in grey or light and dark grey camouflage and were recognisable by their refuelling probe. These RF-4Cs were the only aircraft in the EdA that could be refuelled by flight-boom and the probe-and-drogue method.

The CR.12s were only in service until October 2002, when the last of their kind was retired. All RF-4C aircraft were deployed to 123 Escuadrón 'Titán' within Ala 12 and were used in the reconnaissance role.

Above: When an F-4C takes off, everybody can hear it roar. This aircraft from Ala 12 was eventually replaced with the McDonnell Douglas F/A-18A Hornet.

Right: At the same time the EdA purchased the F-4Cs; it also purchased three Boeing KC-97Ls, which were deployed to 123 Escuadrón in Torrejón de Ardoz. These strategic tanker aircraft enabled the EdA to increase the range of their fighters. This picture was taken in 1974, at Villacisneros, Spanish Sahara.

Between November 1970 and July 1972, the EdA purchased two Piper PA-31 Navajos. They were deployed to 744 Escuadrón from the Escuela de Polimotores. In 1985, the EdA bought a PA-31P-425 type. The aircraft in the picture was parked at Cuatro Vientos AB.

Left: Between the end of 1982 and the beginning of 1983, the AS332 Super Puma helicopters from Aerospatiale (now Eurocopter) arrived in Spain. The 12 AS332s were intended mainly for an SAR role, although two of them acted as VIP transport, and soon two more were purchased.

Below: A superb picture of an AS332 with SAR markings. This helicopter has helped in many search and rescue missions during its years flying with 802 and 803 Escuadrones. The helicopter in this picture is waiting to take off from Alcantarilla AB.

Without a doubt, the Lockheed C-130 Hercules has been invaluable to the EdA. It has flown all over the world, bearing the Spanish cockades, completing humanitarian (Guatemala, Nicaragua, Algeria, Mexico, Colombia, Cameroon, etc) and peace missions (Namibia, Iraq, Somalia, Bosnia-Herzegovina, Ruanda, Albania, Kosovo, etc). The career of the C-130 was a great success; it were officially withdrawn from active service on 31 December 2020.

Right: The C-130 is pictured bearing its usual tricoloured livery. This was the usual look of the aircraft for many years before they were painted grey in the mid-1990s. These aircraft were deployed to Ala 31, which became 31 Grupo in 1989.

Below: This C-130 shows the new grey livery used from the mid-1990s. These aircraft had been the backbone of the transport branch of the EdA for 46 years. Today, the Airbus A400M Atlas has fully replaced them within Ala 31 (which replaced the 31 Grupo during the summer of 1999).

There are many variants of the C-212. One version, known as EC-212, is intended for ELINT (electronic intelligence) and ECM (electronic countermeasure) missions. The special feature of this aircraft is that it has a radome on its nose.

Above: In this picture, an EC-212, painted in white and grey livery, is ready to land.

Left: The C-212 series 200 produced an aircraft perfect for SAR missions. The radome on its nose contains an AN/APS 128 seeking radar. The EdA purchased seven of these aircraft. Note the special livery of the Spanish flag and the yellow band.

Today, 17 venerable C-212s, deployed to the 721 Escuadrón, still fly for the EdA in transport and paratrooper-launching roles. This aircraft has the modern grey livery typical in the current EdA.

In 2002, the EdA took part in the *Red Flag* exercise with a Mirage F-1M from Ala 14. This Mirage F-1M is pictured refuelling as it flies to Alaska. The refuelling probe was typical of the 22 F-1EE (single-seat) and two F-1BE (two-seat) versions of this aircraft, which the EdA purchased to deploy to 462 Escuadrón and received between February 1982 and April 1983.

The EdA was very satisfied with the F-1s, and so it decided to purchase some pre-used versions. Ten F-1EDA and two F-1DDA from the Qatar Air Force were delivered between 1994 and 1997. Initially, the aircraft kept the two-toned livery of their previous owner, but they were later painted in grey.

When the first F-1s arrived in Spain they were painted in a lizard scheme, as can be seen on these aircraft from Ala 14, pictured lined-up at Son San Juan Airport in November 1981.

Left: As time went on, the lizard scheme livery of the F-1s was replaced with NATO light grey. In this picture, three F-1Ms from Ala 14 are flying over La Mancha, near Los Llanos AB. Note the low visibility St Andrew's Cross and Ala 14 emblem. Although it cannot be seen in this picture, these aircraft featured a fake cockpit painted on the underside, designed to trick the enemy.

Below: In the mid-1990s, the F-1s underwent a modernisation programme and 51 single-seaters and four two-seaters were upgraded to M version. Pictured here is a Mirage F-1 during its last years flying for the EdA; it is painted in light grey livery, with a low visibility emblem and St Andrew's Cross.

A beautiful C-101 belonging to the Patrulla Águila (the Spanish aerobatic team) which was formed on 4 June 1985 by teachers from the AGA. Note the spectacular livery of silver, red and yellow, resembling the Spanish flag.

From 1965 to 1985: Improving the Breed

Right: Ninety-two C-101s have been delivered to the EdA – four prototypes and 88 EB series. All Spanish pilots will have had many flight hours with this handsome aircraft. As is standard for trainer aircraft, the wing tips, nose and tail have been painted red. The man in the rear seat is the man responsible for almost all the pictures of this book, Juan Arráez.

Below: Since the first C-101 was delivered to the EdA, it has primarily been deployed to the AGA. For many years these aircraft have excelled in their duties, but very soon they will have to be replaced. Note the AGA emblem painted on the tail of this aircraft.

Two C-101EBs from 794 Escuadrón. This escuadrón is the same as the Patrulla Águila and has seven C-101s.

793 Escuadrón is intended for basic flight training in the AGA. Fifty-four C-101 EBs have been delivered during the years to 793 and 794 Escuadrones. The young pilots need to fulfil 48 flight hours in the E.26 to be able to fly the C-101.

Several C-101s from AGA flying in formation. It is noteworthy that only the aircraft in the background has a pointed tip, while the other tips are rounded. The reason for this is that the first C-101s were fitted with a silver, rounded tip, however, shortly after, it was found that ice formed dangerously around this style of tip, and the risk of the ice falling into the engine's air intake was high.

The EdA has decided to purchase up to 18 CASA C-295s, and the first of these entered service in Ala 35 in 2001. Currently, there are 13 flying for the EdA. This modern transport aircraft will be the perfect partner of the bigger Airbus A400M Atlas.

From 1965 to 1985: Improving the Breed

When the DHC-4 became obsolete, it was replaced by the Spanish-built CASA CN-235, which arrived in July 1991. These aircraft have had roles in transport, the Grupo de Escuelas Matacán (GRUEMA), SAR, CECAF and maritime surveillance.

In 1971, the EdA purchased several Canadair CL-215 amphibious aircraft, and over the following years, the total number of Canadairs increased to 14. Some of those now in service belong to the improved CL-215T type, like the aircraft in this picture.

A beautiful picture of a CN-235 in its green, brown and sand livery. Note the rear door, which is opened for a show. The aircraft in this picture belonged to Ala 35 and was intended for a transport role.

F-5s lined up and displaying the different camouflage schemes they had.

These two F-5s are flying in a fighter role; this is evident because of the aircraft number painted on the tail, which begins with 'C', the letter used to denote a fighter. They have a silver livery, as was the standard during the 1970s.

Chapter 4
From 1985 to 2000: Sharpening the Claws

As the late 1980s approached, EdA fighters were getting older and, once again, the need for replacement aircraft was looming.

Fighters

An early candidate for replacement was the faithful F-4C. Nearly 20 years after the first F-4 arrived in Spain, 18 reconnaissance versions of the aircraft, called RF-4Cs (CR.12), joined the EdA. As with the Spanish-owned F-4s before them, these aircraft were not new at time of purchase and had flown previously for the 165 TRS Kentucky Air National Guard. These aircraft were deployed to 123 Escuadrón 'Titán', based at Torrejón de Ardoz AB. The original F-4Cs flew reliably within the EdA until 1991, while the reconnaissance versions fulfilled their duties until 2002 when they were also retired.

However, even before the official retirement of the F-4s, the EdA was anxious to find a new fighter. In 1982, Spain joined NATO, and the close relationship between Spain and the US was reaffirmed. This was undoubtedly a reason behind the purchase of a replacement American-built fighter, the McDonnell Douglas F-18 Hornet. After studying the performance of several fighters in the FACA (Futuro Avión de Caza y Ataque/Future Fighter and Attack Aircraft) programme, the F-18 was officially chosen in May 1983 and 72 newly built aircraft were purchased. Specifically, there were 60 McDonnell EF-18As (single-seater) and 12 EF-18Bs (CE.15). This aircraft provided the EdA with a fighter with high attack and air defence capacities, with great efficiency and guaranteed availability. The first F-18s arrived at Zaragoza AB in July 1986 to join Ala 15, created on 16 December 1985, which included 151 Escuadrón 'Toro', 154 Escuadrón 'Marte' and 153 Escuadrón 'Ebro'. In 1989, Ala 12 received the F-18 as a replacement for its F-4Cs. With the arrival of the F-18s, a new unit was formed to join 121 and 122 Escuadrones: 123 Escuadrón 'Titán'.

A beautiful shot of an F-18 from Ala 46. This aircraft has a special livery painted on the tail to commemorate the 50th anniversary of Ala 46; the chosen design represents all the aircraft that have flown in the Ala. Ala 46, together with Ala 12 and Ala 15, are the three main units that fly F-18s. This unit is based at Gando AB, in the Canary Islands, and is tasked with protecting the southwestern flank of Spain.

The Spanish F-18s were improved to F/A-18As and F/A-18Bs, and, in October 1994, the EdA acquired 24 F/A-18As from US Navy surplus, which arrived in Spain between 1996 and 1998. In 1996, the new F-18s were sent to Ala 21, but soon after they were redeployed to 462 Escuadrón.

The Spanish F-18s were modernised by CASA; the improved version was named EF-18M and, by 31 October 2009, most aircraft had undergone this development process. However, despite these attempts at modernisation, the F-18 fleet is getting older, and will inevitably need to be replaced by a new aircraft very soon, but the Spanish budget for the EdA is currently a subject of concern.

In November 1992, the F-5As and F-5RFs from Ala 21 were redeployed to the Ala 23. Today, 22 upgraded F-5s, named SF-5M, still fly for the EdA as advanced combat trainers deployed to 231 Escuadrón 'Patas Negras' and 232 Escuadrón, based at Talavera la Real AB.

Additional Aircraft

The Cessna Citation V entered service within the EdA in December 1992. Today, three Cessna C560 Citation Vs carry out aerial photographic missions from Getafe AB.

VIP Transport

In 1978, the EdA purchased a Douglas DC-8, intended for long-range VIP transport. The aircraft was a success, and a second DC-8 was acquired in June 1980. However, it was not long before the DC-8 started to be replaced by the Boeing 707, with the first substitutions being made as soon as March 1988, owing to technical problems in both Douglas aircraft. The EdA operated up to four Boeing 707s from 1988 until 2017. The first 707 replaced the DC-8 from 401 Escuadrón in a VIP transport role. The other 707s acquired were adapted as VIP transports, freighters and used in medical evacuation roles. Later, these aircraft were deployed to 45 Grupo.

Only one Dassault Falcon 50 was ever purchased by the EdA, flying from May 1982 until June 2003. Its main role was as a long-range VIP transport. It was deployed to 401 Escuadrón.

Since April 1988, five Dassault Falcon 900B aircraft were operated by the EdA. They were designed for long-range VIP transport and humanitarian aid missions. They still fly for the EdA today and are deployed to 45 Grupo.

A superb shot of an F-18M, but this time from Ala 15. The aircraft has a special livery intended for Tiger Meet 2016, which took place in Zaragoza and was organised to commemorate the 30th anniversary of Ala 15. Just behind this F-18M, several Eurofighter EF2000 Typhoons can be seen.

Two Airbus A310s were acquired and entered service in 2003. In the same way as the 900Bs, these aircraft were deployed to 45 Grupo and continue to fulfil a VIP transport role.

Helicopters

In order to improve its helicopter training programmes, the EdA acquired two different types: the Sikorsky S-76 and the Eurocopter EC120 Colibri. The S-76 entered service in 1991 and continues to be used as a trainer for instrument flights. Eight S-76s were purchased and all are deployed to Ala 78, based at Armilla AB.

The old 296s from the Escuela de Helicópteros were replaced by new EC120 aircraft, which were purchased on 22 December 1999 and operated in a training role. As a credit to its good performance, five of these helicopters are still used by the EdA's helicopter aerobatic team, ASPA.

Right: A close-up shot of the rear part of an F-18 from Ala 12, displaying a special livery intended to commemorate the 50th anniversary of the Ala. We can see the shapes of the aircraft that flew in the Ala, including the F-4C, the F-86, the F-104 and, of course, the F-18.

Below: A two-seater EF-18M with a special anniversary livery. The emblem and motto on this aircraft signify that it belongs to Ala 12.

Spain has purchased two Airbus A310-304s, which were deployed to 45 Grupo. This aircraft provides maximum flexibility for both medium- and long-range routes with a high degree of reliability, and so it is no surprise it has been chosen by the EdA to work in a VIP transport role.

Above: The Airbus A310 in this picture is ready for take-off at San José Airport, Costa Rica, surely flying off to complete a VIP transport mission. Similar to the DC-8s, the Airbus A310 has an all-white livery, only divided by a central red line.

Left: The EdA operated up to four Boeing 707s between 1988 and 2017. These aircraft were intended for VIP transport and as a replacement for the DC-8s, but they were quickly replaced by the Airbus A310 because of technical problems.

Right: A Douglas DC-8 is pictured flying over Gando. Note the all-white livery, only divided by a central red line. This aircraft was purchased in 1978 for long-range VIP transport.

Below: Due to the success of the first DC-8 within the EdA, a second DC-8 was acquired in June 1980. However, owing to several technical problems, the EdA was forced to replace the aircraft early, and opted for a used 20-year-old Boeing 707. A lonely C-101 is parked close to the DC-8 at San Javier AB.

This Cessna Citation V is pictured resting at its base. Three aircraft of this type were purchased by the EdA in December 1992, and today they are deployed to CECAF, together with several C90s and a CN-235.

One of the Cessna C560 Citation Vs of the EdA. This aircraft carries out mainly aerial photographic missions from its AB at Getafe. Note the usual livery of this C560, which has the upper side of the fuselage painted in white and the lower side in light grey, divided by a central black line.

Left: The Sikorsky S-76 helicopter entered service in 1991 and is used as a trainer for new pilots, primarily for instrument flights. All of these helicopters are deployed to Ala 78, based at Armilla AB, Granada.

Below: Although the eight S-76s purchased by the EdA are used as trainers, the high capacity and high manoeuvrability of this aircraft mean that is also used for VIP transport or SAR roles in other countries.

The S-76 is much appreciated by its crew because of its high performance. In this image, an S-76 is on show at Sevilla during the Spanish Armed Forces Day in 2019. Note its light grey livery and the red band on the tail, as was usual for trainer aircraft. (Eduardo Manuel Gil Martínez)

A beautiful shot of a Spanish Dassault Falcon 20, which started to fly for the EdA in 1970 and retired in 2005. The main role of the Falcon 20 was to provide VIP transport for Spanish officials. Note the large St Andrew's Cross on the tail and the white livery divided by a central red line, as was usual for VIP transport aircraft.

This Falcon 20 has light grey livery and a small St Andrew's Cross on the upper tail, as was common in its last years of service. Fifty-five of these aircraft were purchased by the EdA and deployed to 45 Grupo.

Only one Dassault Falcon 50 was ever purchased by the EdA, flying from May 1982 until June 2003. Its main role was as a long-range VIP transport aircraft and it was deployed to 401 Escuadrón.

Left: Another shot of the same Falcon 50 as previously pictured. It has retained the white livery with the central red line of a VIP transport aircraft, but it now has a different UN (45-20 has become 401-09) and has a Spanish flag painted in front of the tail.

Below: The Eurocopter EC120 'Colibri' has modern avionics, high manoeuvrability and can transport five crew; these factors make this helicopter perfect for both basic and tactical training. On 22 December 1999, 19 EC120s were purchased and deployed to Ala 78.

Right: EC120s are used by the EdA's helicopter aerobatic team, ASPA. Note the Spanish flag over the light grey fuselage and the red band on the tail. (Eduardo Manuel Gil Martínez)

Below: Five EC120s form the helicopter aerobatic team, ASPA. In this picture, three helicopters are performing difficult manoeuvres at high speed. If you see their show, you will never forget its beauty, elegance and difficulty.

An F-18 from Ala 15 is ready to take off. During the Yugoslav Wars, the main Spanish contribution to NATO missions was these aircraft.

One of the F/A-18A+ belonging to Ala 15 shows the typical air superiority grey livery. Although these fighters have many years of service behind them, they remain a very important and reliable aircraft for the EdA.

Between 1996 and 1989, 24 F-18s were purchased by the EdA from the US Navy. When they arrived, they were painted in a tricoloured camouflage that resembled Soviet fighter aircraft.

Chapter 5

From 2001 until Today: The Most Powerful EdA of all Time?

The EdA's dedication to the continuous improvement of its aircraft has established it as one of the most dominant air forces in the world.

Fighters

The most modern fighter to join to the EdA is the Eurofighter EF2000 Typhoon. The EdA acquired 73 EF2000s between 2004 and 2020, including single-seaters named C.16 and twin-seaters named CE.16. The first EF2000s to arrive in Spain in October 2004 were deployed to Ala 11, which comprised 111 Escuadrón and 113 Escuadrón. In April 2021, the aircraft was deployed to Ala 14 to replace the venerable F-1s. In response to the success of the EF2000s, the EdA has decided to purchase 20 new ones, which should be delivered in 2025–30, in order to assist with Proyecto *Halcón* (Project *Falcon*).

Transport

The newest transport aircraft for the EdA are the CASA C-295 and the Airbus A400M Atlas. The C-295 is a new, Spanish-built medium transport aircraft, based on the CN-235, that is the ideal complement to the A400M. The EdA intends to acquire up to 18 C-295s in total, and the first one entered service in Ala 35 in 2001. Currently, there are 13 flying for the EdA.

A C-295 transport aircraft from Ala 35 shows us its modern and beautiful profile as it flies. It is the ideal complement to the A400M. Currently, 13 C-295s are flying for the EdA from their base at Getafe.

One of the NHIndustries NH90 belonging to the EdA is pictured mid-flight. This helicopter is intended for an SAR role, although it can fulfil other duties such as Special Air Operations or Personal Recovery in a hostile environment. (Pablo Rada via Juan Arráez)

The first A400M was delivered to the EdA on 17 November 2016, and it was incorporated into Ala 31 on 1 December 2016, where it coexisted with the C-130 for four years. Ala 31 currently has nine A400Ms, and they have provided a vast improvement to the EdA's transport capabilities. Over the next few years, the EdA will have 14 A400Ms in service. The tanker version of the A400M is named TK.23.

Alongside the modern C-295s and A400Ms, 17 C-212s continue to work as transport craft with 721 Escuadrón, based in Alcantarilla AB. Ten CN-235s also remain deployed at the Centro Cartográfico y Fotográfico (CECAF, Cartographic and Photographic Centre) and eight to the Grupo de Escuelas Matacán (GRUEMA), serving as tactical transport and training craft.

ASW and Maritime Patrol

Maritime surveillance capabilities have been improved by the addition of the CN-235. Initially, there were eight of these aircraft within the EdA, which were split up as follows: three deployed to 801 Escuadrón; two deployed to 802 Escuadrón, based at Gando AB; and three to 803 Escuadrón, based at Getafe AB. Today, all CN-235s belong to Ala 37 and operate in an SAR role.

Additional Aircraft

CECAF oversees mapping and photographic tasks within the EdA. In addition to the aforementioned CN-235s, it also continues to utilise Citation Vs and C90s. The Citation Vs are used as VIP transport from time to time.

The first Spanish A400M belonging to Ala 31 is ready to take off. Note the refuelling probe on the port side of the nose. Soon, Spain will have 14 A400Ms in service.

As discussed above, C90s entered service in 1974 and continue to fly for the EdA. However, since July 2003, they have also been redeployed to the 42 Grupo as VIP transports and 409 Escuadrón as a VIP transport and liaison aircraft.

Helicopters

The EdA has recently purchased 12 NHIndustries NH90 helicopters, although, as of 2020, only two had been delivered. At the time of writing, four more NH90s are scheduled to be delivered. Then, in 2022, a second batch of 23 brand new NH90s will start to be delivered. The NH90 is a medium-size, twin-engine multi-purpose helicopter and is deployed to 803 Escuadrón from Ala 48, based at Cuatro Vientos AB.

One of the last acquisitions made by the EdA at the time of writing is the General Atomics MQ-9 Predator B. This Unmanned Aerial Vehicle (UAV) was considered necessary to achieve better effectiveness and efficiency in the EdA. Four MQ-9 Block 5s arrived in Spain in 2019–20, with two remote control stations, and are intended mainly for a reconnaissance role. These UAVs were deployed to 223 Escuadrón, belonging to Ala 23, based at Talavera la Real AB.

Training

The EdA has plans to replace its C-101s and F-5s in the near future. Although it is not decided, it is possible that the replacement aircraft would be the AFJT (Airbus Future Jet Trainer), which could be available in 2026.

At present, the new PC-21 is going to replace the C-101s for basic training. Twenty-four PC-21s will be delivered in the next two years (2021–23), while the E.26 continues as an elementary trainer. The first two PC-21s arrived in Spain on 14 September 2021 and will start their training role in 2022.

Current Numbers

At the time of writing, the number of EdA aircraft according to the main different types is as follows:

Fighters: 70 EF2000s, 86 F/A-18s (53 EF-18Ms, 12 EF-18BMs) and 20 F/A-18A+s.
Patrol: Two/three P-3Ms.
Transport: Nine A400Ms, 13 C-295s, eight CN-235s, 17 C-212s.
Trainers: 19 F-5BMs, 62 C-101s, 36 T-35s, four AS 332s, two AS532s and 16 F33Cs.
VIP transport: Two A 310s, five Falcon 900s and three Citation Vs.
Helicopters: 15 EC 120s, eight S-76Cs, 13 AS 332s and two NH90s.
UAVs: Four MQ-9 Block 5s.

The NH90 is the first production helicopter to feature entirely fly-by-wire flight controls, which have greatly enhanced mission capabilities. In the EdA, this helicopter is based at Cuatro Vientos AB and belongs to 803 Escuadrón from Ala 48. (Pablo Rada via Juan Arráez)

Above: As Airbus advertising says, the A400M is a superb aircraft that can carry any load and deliver it even to the smallest and most unprepared airstrips. When acting as a tanker, the A400M has a basic fuel capacity of 63,500 litres and even more with additional tanks.

Left: An underside view of an A400M as it flies trials over San Javier. Thanks to this big aircraft, the capabilities of the transport branch of the EdA have dramatically increased. Today, this grand aircraft is considered to be one of the best military heavy transport aircraft in the world.

The NH90 is a modern helicopter developed in response to NATO requirements for a modern battlefield helicopter that is capable of all-weather day and night operations. This aircraft type possesses the lowest radar signature in its class, principally due to its diamond-shaped composite fuselage. (Pablo Rada via Juan Arráez)

From 2001 until Today: The Most Powerful EdA of all Time?

The EF2000 is the most modern fighter in the EdA and the current backbone of Spanish air defence. This aircraft's UN indicates that it belongs to the Ala 11 and is based at Morón AB.

Right: The single-seat version of EF2000 is a twin-engine, canard delta wing, multi-role fighter and it is considered to be the second-best fighter in the world, only surpassed by the Lockheed Martin F-22 'Raptor'.

Below: A beautiful picture of an EF2000, belonging to Ala 14 and based at Los Llanos AB. We can see the typical Spanish markings: cockades, UN, aircraft number, the St Andrew's Cross and Ala 14 emblem painted in the tail. This emblem depicts the literary character of Don Quixote hailing three aircraft, chosen because of the character's connection to the local area.

Left: A pair of EF2000s from Ala 14 are pictured during a flight over La Mancha. These aircraft had been sent to the Baltic countries and to Romania under NATO command in order to control the airspace of these countries. Today, Spain has about 70 of these aircraft in service.

Below: This single-seater EF2000 from Ala 11 is ready to land, displaying the light grey livery typical in the EdA now. The first sales contracts of this aircraft were signed in January 1998.

Thanks to this superb fighter, the EdA currently has a high defensive and offensive capacity. Despite this, the EdA has decided to further increase its power with 20 new EF2000s, which should be purchased in 2025–30 and will replace the F-18s based at Gando AB.

From 2001 until Today: The Most Powerful EdA of all Time?

Right: An EF2000 from Ala 11 is receiving some attention in a maintenance hangar as it takes shelter from the hot Spanish sun. This aircraft type and its crews have made Alas 11 and 14 some of the most powerful units among the NATO countries.

Below: One of the aircraft that has played, and continues to play, a very important role in the EdA is the F-5. Seventy F-5s were licence-built in Spain, and all were in service by 1971. After many improvements and changing of roles, this superb aircraft is still used as an advanced combat trainer for Spanish pilots. In this picture, we can see one of the 22 upgraded F-5s, named SF-5M, still deployed to Ala 23.

Above: A render of the candidate thought to be replacing the C-101s and F-5s as an advanced trainer in 2027–28: the Airbus Future Jet Trainer (AFJT). It is believed that Spain will purchase around 50–55 aircraft from Airbus Defence. (Airbus Defence)

Right: If the AFJT comes to life, it will be a perfect substitute for the old Spanish advanced trainers, because it will be highly operational, safe and reliable. In the render, the aircraft is painted in grey livery with Spanish markings. (Airbus Defence)

Above: The first flight of a Spanish-owned Pilatus PC-21 took place in Stans, Switzerland. The aircraft has a grey, yellow and red livery, although it still bears a little Swiss flag on the tail. This PC-21 is the future of the EdA's elementary trainer aircraft. (Pilatus)

Left: This PC-21 is complete with all the Spanish markings: Spanish cockades in six positions, the St Andrew's Cross and the AGA's emblem on the tail. Note the aircraft number, E.27-01, which tells us that this is the first of the PC-21s to fly for the EdA. (Pilatus)

Below: First flight of the General Atomics MQ-9 Reaper, also known as Predator B, within the EdA. Spain has purchased four MQ-9s, which have been deployed to 223 Escuadrón and three Ground Control Stations. (Ministerio de Defensa de España)

The Spanish MQ-9s have a light grey livery and bear the Spanish cockade in six positions and the St Andrew's Cross on the tail. These aircraft are based at Talavera la Real AB, but the intention is to fly them temporarily from Lanzarote. (Ministerio de Defensa de España)

A beautiful picture of one of the last F-5s flying for the EdA. There is a strong possibility that these F-5Ms will be replaced in their advanced trainer role by the AFJT.

One of the EF2000s deployed to the Baltics for the air defence of Estonia, Lithuania and Latvia. Since 2004, several Spanish aircraft detachments had been deployed as part of the Baltic Air Police.

Appendices

Appendix 1

AIRCRAFT NAMES

The Spanish names for their aircraft are sorted by role, based on the work of José Luis González Serrano. The naming system has been changed several times since 1939, however the letters used since 1945 are as follows:

Letter	Mission	In Spanish
A	Attack	Ataque
B	Bomber	Bombardero
C	Fighter	Caza
D	Rescue	Salvamento
E	School	Escuela
K	Tanker	Cisterna
L	Liaison	Enlace
M	Electronic warfare	Guerra electrónica
P	Patrol	Patrulla
R	Reconnaissance	Reconocimiento
S	Anti-submarine	Antisubmarino
T	Transport	Transporte
U	Utility	Utilitario
V	Glider	Velero
X	Prototype	Prototipo
Z	Helicopter	Helicóptero

The tables below show the different names within the EdA of the helicopters, fighters, bombers, transport, seaplanes, gliders and trainer aircraft.

Opposite below: Even after the SCW came to an end, the EdA maintained the period insignias, albeit with some modifications. Although it was already used during the SCW on some occasions, the EdA widened the use of the Spanish cockade. This consisted of a circle with the three colours of the Spanish national flag: red in the outside ring, yellow in the middle ring and red in the inner disc. The cockade is placed in four positions on the wings and sometimes on both sides of the fuselage. Here, we can see an He 111 without and an He 70 with the Spanish cockade.

Appendices

Fighters

Aircraft	Spanish Nickname	1939–45	1945–54	1954–78	1978–Present
Fiat G.50		1	C.6		
Heinkel He 51		2	A.1		
Fiat CR.32	Chirri	3	C.1	C.1	
Hispano HA-132L		3	C.1	C.1	
Heinkel He 112 B		5	C.3		
Messerschmitt Bf 109 B, C, D	Bipala (Two bladed)	6	C.4 → C.4B		
Messerschmitt Bf 109 E	Tripala (Three bladed)	6	C.5 → C.4E	C.4E	
Romeo Ro.41		7	C.7		
Polikarpov I-15	Curtiss	2W, 8	A.4	A.4	
Arado Ar 68		9	C.11		
Polikarpov I-16	Rata (Rat)	1W	C.8		
Polikarpov I-15bis		2W	C.9		
Hawker Fury		4W	C.2		

Aircraft	Spanish Nickname	1939-45	1945-54	1954-78	1978-Present
Grumman G-23		5W	R.6		
Messerschmitt Bf 109 F	Zacuto	6	C.10 → C.4F		
Hispano HA-1109 J, K	Jota		C.12	C.4J	
Hispano HA-1112 M	Buchón (Pouter)			C.4K	
North American F-86F				C.5	
North American T-6D				C.6	AE.6
Lockheed F-104G				C.8	
Lockheed TF-104G				CE.8	
CASA Northrop F-5A				C.9	A.9
CASA Northrop RF-5A				CR.9	AR.9
CASA Northrop F-5B				CE.9	AE.9
Hispano HA-200A	Saeta			C.10A	A.10A
Hispano HA-200D	Saeta			C.10B	A.10B
Hispano HA-220	Super Saeta			C.10C	A.10C
Dassault Mirage III EE	Plancheta (Iron)				C.11
Dassault Mirage III DE	Plancheta				CE.11
McDonnell F-4C					C.12
McDonnell RF-4C					CR.12
Mirage F-1 (single seat)					C.14
Mirage F-1 (two-seat)					CE.14
McDonnell EF-18A					C.15
McDonnell EF-18B					CE.15
Eurofighter EFA-2000A					C.16
Eurofighter EFA-2000B					CE.16

After the SCW, the winning side used round, black badges on both sides of their aircraft's fuselage to avoid cases of mistaken identity. Notice the black badge painted on this He 59.

Bombers

Aircraft	Spanish Nickname	1939–45	1945–54	1954–78	1978–Present
Savoia S.81		21	T.1		
Junkers Ju 52/3m		22	T.2 (BMW engine)	T.2B (ENMASA BETA engine)	
Fiat Br.20		23	B.3		
Henschel Hs 123	Angelito	24	BV.1	BV.1	
Heinkel He 111 B, E	Pedro	25	B.2 (B, E)	B.2	
Junkers Ju 86 D		26	B.4		
Dornier Do 17 E, F, P	Bacalao (Cod)	27	R.3		
Savoia S.79		28	B.1		
CASA C-2.111 A, C			B.7	B.2H/B.2HR	
CASA C-2.111 B, D			B.7	B.2I/BR.2I	
Tupolev SB	Katiuska	20W	B.5		
Junkers Ju 88 A, C, D		29	B.6		

The EdA's aircraft have a St Andrew's Cross painted on a white background on the tail. From the late 1970s, the size of the emblem was reduced. Notice this C-101 bearing the cross.

This HA-220 helps to demonstrate how the aircraft numbering system worked within the EdA. For example: 'C' signifies that the aircraft is a fighter; '10C' shows the aircraft type, in this case an HA-220, with a letter after the number, indicating that it is an improved variant; '107' represents the serial number of each aircraft in the HA-220 series flying for the EdA.

Spanish Air Force Aircraft: 1939–2021

Transport

Aircraft	1939–45	1945–54	1954–78	1978–Present
Douglas DC-1 and DC-2	42			
Ford 4T Trimotor	42			
Fokker F-XII	45			
CASA C-352L	22	T.2	T.2 T.2B	
Douglas C-47		T.3	T.3	
Focke Wulf Fw 200C		T.4		
Lockheed 18 Lodestar		T.4		
Douglas C-54			T.4	
CASA C-201		XT.5	T.5	
CASA C-202			XT.6B	
CASA C-207A and C			T.7	
CASA C-2.111 E, F, G, H			B.2H T.8 T.8B	
Fairchild C-119 F			T.9	
De Havilland DHC-4			T.9	T.9
Lockheed C-130H			T.10	T.10
Lockheed KC-130H			TK.10	TK.10
AMD Falcon 20			T.11 TM.11	TM.11
CASA C-212 /several versions)				T.12 TE.12 TR.12 TM.12
Convair CV-440			T.14	
Douglas DC-8-53				T.15
AMD Falcon 50				T.16
Boeing B-707				T.17 TK.17 TM.17
AMD Falcon 900				T.18
CASA CN-235				T.19 TR.19
Cessna C-560				TR.20 TM.20
CASA C-295				T.21
Airbus A-310				T.22
Airbus A-400M				T.23 TK.23
Boeing KC-97L			TK.1	

This F-4 can help explain the UNs painted on Spanish aircraft. For example: The first number, '122', identifies the unit that the aircraft belongs to, which in this case is 122 Escuadrón; the number '15' is the aircraft's order number into its unit, however, this is not necessarily chronological.

Trainers

Aircraft	1939–45	1945–54	1954–78	1978–Present
Caproni AP.1	32	ES.4		
Bücker Bü 131	33	EE.3	E.3B/E.3	E.3B
CASA C-1131 E/H	-/33	EE.3	E.3B/E.3	E.3B
Bücker Bü 133	35	ES.1	E.1	
CASA C-1133L	35	ES.1	E.1/E.1B	
Arado Ar 66C	36	ES.7		
Fiat Cr.30	37			
Romeo Ro.41 (two seat)	37	ES.3		
Gotha Go 145	38	ES.2		
CASA C-1145L	38	ES.2		
Fiat Cr.32 (two seat)	3	ES.5		
Hispano HA-132L	3	ES.5		
Hispano HS-42	39	ES.6	E.6	
Hispano HS-43		ES.9	XE.6	
Huarte Mendicoa HM-5		ES.8	E.8	
DH 60 G III	30	EE.1		
DH 82A	30	EE.1		
Hispano E-30	30	EE.2		
Huarte Mendicoa HM-1	30	EE.4	E.4	
Huarte Mendicoa HM-9		EE.5	E.5	
AISA I-115		EE.6	E.6	
Hispano HA-200			E.14	E.14A
Lockheed T-33A			E.15	E.15
North American T-6G			E.16	E.16
Beechcraft T-34 A/B			E.17	E.17
Piper PA-31			E.18	E.18
Piper PA-23E			E.19	E.19
Beechcraft B-55			E.20	E.20
Hispano MBB-323			E.21	
Beechcraft C-90			E.22	E.22/U.22
Beechcraft A-100			E.23	
Beechcraft F-33C			E.24/E.24A	E.24A
Beechcraft F-33A			E.24B	
CASA C-101 EB				E.25
CASA/ENAER T-35				E.26
Pilatus PC-21				E.27

Seaplanes

Aircraft	Spanish Nickname	1939–45	1945–54	1954–78	1978–Present
CASA Dornier Wal		70	HR.1	HR.1	
Heinkel He 60		60	HR.2		
CANT Z.501		62			
Arado Ar 95		64	HR.3		
Heinkel He 114		61	HR.4		
Dornier Do 24T-3		65	HR.5	HR.5/HD.5	
Fairey Swordfish			HR.6		
Romeo Ro.43			HR.7		
Heinkel He 59		71			
CANT Z.506		73			
Consolidated OA-10		66	DR.1	DR.1	
Grumman HU-16A SAR				AD.1	D.1A
Grumman HU-16B SAR				AD.1B	D.1B
Grumman HU-16B ASW				AN.1/AN.1A	S.1A
Grumman HU-16B ASW (purchased from Norway)				AN.1B	S.1B
Canadair CL-215/CL-215T	Botijo			UD.13	UD.13
Canadair CL-415	Botijo				UD.14

The UN is always painted on both sides of the forward or rear fuselage, and the two numbers are either separated by a hyphen or a Spanish cockade. As can be seen on this CN-235.

Helicopters

Aircraft	Spanish Nickname	1939–45	1945–54	1954–78	1978–Present
Sikorsky H-19A				Z.1C	
Sikorsky H-19B				Z.1	
Sikorsky H-19D				Z.1A	
Westland Whirlwind Mk.52				Z.1B/ZD.1B	
Aerotécnica AC-12 Pepo				Z.2	
Hiller OH-23C				Z.6	
Agusta Bell AB-47G-2				Z.7	HE.7
Agusta Bell AB-47G-3B				Z.7B	HE.7B
Agusta Bell AB-47J-3B1				Z.11	HE.11
Bell OH-13H				Z.7A	HE.7A
Agusta Bell AB-205				Z.10	HD.10A/HE.10A
Bell UH-1				Z.10B	HD.10B/HE.10B
Agusta Bell AB-206 Jet Ranger				Z.12	HD.12
Sud Aviation SA-319B Alouette III				Z.16	HD.16
Sud Aviation SA-330C/H/J/L Puma				Z.19	HT.19/HD.19
Hughes H-269C					HE.20
Aerospatiale AS-332B Super Puma					HT.21/HD.21
Aerospatiale AS-332L Super Puma					HT.21A
Sikorsky S-76C					HE.24
Eurocopter EC-120B Colibri					HE.25
Eurocopter AS-532L Cougar					HD.27
Eurocopter NH 90	Lobo (Wolf)				HD.28

Gliders

Aircraft	1939–45	1945–54	1954–78	1978–Present
Schneider (DFS) Grünau Baby		V.1		
Hans Jacobs (DFS) Kranich III		V.2		
Jacobs Schweyer (DFS) Weihe		V.3		

Another example of a UN, this time on a twin-seat EF2000 from Ala 11. (Public domain)

Appendix 2

NATIONAL MARKINGS, AIRCRAFT NUMBER AND UNITY NUMBER

National Markings

In Spain, after the SCW, the aircraft of the winning side used round black badges on both sides of the fuselage, although sometimes the Falange Española de las JONS (Spanish Phalanx of the Councils of the National Syndicalist Offensive) symbol of the yoke and arrows was added to the interior. The black badges were painted from 8 August 1936, replacing the tricolour badge of red-yellow-red in order to avoid confusion with the insignia of the Republican side, which used red-yellow-purple. On the wings, round black or black diagonal cross insignia were commonly used. The tail featured a St Andrew's Cross, also called the Cross of Burgundy, which is similar to the cross used in the Bulgarian Air Force during World War Two when it was Germany's ally. The cross represents the martyrdom of the apostle,

On this EF2000 we can see the Ala 11 emblem of a bull, with the motto 'Look, luck and to the bull', alongside a Spanish cockade.

and its use in Spain dates back to the time of the marriage of Joan of Castile with Philip the Beautiful and is attached to the Spanish shield.

Usually, the Spanish cockade is placed in six positions up and down wings and on fuselage, and the St Andrew's Cross is on the rudder. Both the Spanish cockade and St Andrew's Cross continue to be used on Spanish aircraft today.

Aircraft Numbers

Here is a brief explanation of how the EdA assigned numbers to its aircraft. To demonstrate, let us look at the C.16-52:

- C: The first letter designates the aircraft's intended role, in this case C for Caza, or fighter.
- 16: The first number designates the aircraft type. In this case, this is the 16th fighter type in the EdA: the EF2000. Sometimes, after the number, there is a letter to indicate the variant.
- 52: The number after the hyphen is the serial number of each aircraft in the EF2000s series flying for the EdA.

The aircraft number is painted on both sides of the tail.

Unity Number (UN)

The UN is always painted on both sides of the forward or rear fuselage, and it has two numbers separated by a hyphen or, in some cases, the Spanish cockade. For example, the 11-18.

- 11: The first number identifies the unit that the aircraft belongs to – sometimes this is the Ala, but it can also be an Escuadrón or Grupo, because there are independent Grupos and Escuadrones that do not belong to an Ala. However, there are so many variants that it is difficult to establish a general rule, and this text does not aim to attempt to.
- 18: This is the aircraft order number into its unit, which is not necessarily correlative or chronological.

If these numbers are separated by the Spanish cockade, they are painted at the rear of the fuselage, but if the aircraft shape does not allow it, both numbers are painted in the forward fuselage and the Spanish cockade in the rear fuselage, after the wings.

Two HA-220s from Morón AB are pictured flying over southern Spain. Both the HA-200 and the HA-220 were big steps for the Spanish aeronautical industries, which proceeded to manufacture other successful aircraft, including the C-212, C-101, CN-235, C-295 and collaborate on international projects such as the EF2000 and the A400M.

Appendix 3

ESCUADRONES RADIO CODENAMES
Here is a compilation of the radio codenames that are used by various escuadrones.

Escuadrón	Radio Codename
211	Gallo
212	Sisón
122	Tenis
123	Titán
231	Patas Negras
232	Mago
151	Toro
153	Ebro
154	Marte
301	Dumbo
141	Chico
142	Dardo
121	Póker
111	Dólar
462	Halcón

Appendix 4

USEFUL TRANSLATIONS

AGA or Academia General del Aire	Air General Academy
Ala/Alas	Wing/Wings
Ala de Caza	Fighter Wing
Ala de Transporte	Transport Wing
CASA (Construcciones Aeronáuticas Sociedad Anónima)	Aeronautical Manufacturing Inc
CECAF or Centro Cartográfico y Fotográfico	Cartographic and Photographic Centre
CLAEX or Centro Logistico de Armamento y Experimentación	Logistic Centre for Armament and Experimentation
Destacamento	Detachment
Ejército del Aire (EdA)	Spanish Air Force (literally Air Army)
Escuadrón (escuadrones)	Squadron (squadrons)
Escuela Elemental	Elementary School
Escuela de Reactores	Jet Training School
Grupo de Estado Mayor	Headquarters
Fuerza Aérea Española	Spanish Air Force

GRUEMA or Grupo de Escuelas Matacán	Matacán School Group
Mando	Command
Mando de Tansporte	Transport Command
Patrulla	Patrol
Regimiento	Regiment
Región Aérea	Air Region
SAR or Servicio Aéreo de Rescate	Search and Rescue Service
Número de Unidad	UN or Unity Number

Appendix 5

WAR AND INTERNATIONAL MISSIONS

It is not the aim of this text to detail the international missions that the EdA has carried out, however, it is still worth writing about some of these matters.

The EdA has taken part in relatively few armed conflicts since it was created, although the ones it has participated in are notable. Aside from direct combat, the EdA has participated in numerous international missions providing humanitarian aid and military surveillance. Here, we will comment briefly on some of these events.

During World War Two, in spite of proclaimed Spanish neutrality, Spanish pilots flew within the Luftwaffe against the USSR. Furthermore, in the same way as with the Army, the Air Force collaborated with the Axis powers, as shown by the authorisations and facilities that were given to Germans and Italians in Spain. However, the most notable contribution from Spain during this time was the Escuadrillas Azules (Blue Squadrons), which flew with the Luftwaffe in order to support Germany against attacks from the Eastern Front.

The five units sent by Spain fought on rotation in the Russian skies between September 1941 and March 1944, with one squadron being relieved by another roughly every six months. The Spanish pilots had the opportunity to fly Bf 109 E7s, Bf 109 F2s, Bf 109 F4s, Focke-Wulf Fw 190 A2s, Fw 190 A3s and Bf 109 G6s.

After the end of World War Two, the EdA had not officially taken part in any war, but in 1957, Spanish aircraft had to defend Spanish territories in Western Africa. After obtaining independence in 1956, Morocco began to express its interest in annexing Spanish possessions in Africa. The Moroccan sultan supported and financed the rebel bands of the Moroccan Army of Liberation against Spanish territory. Following the attacks from Morocco at the end of 1957, Spain entered a war against insurgents to defend their territories in Sidi Ifni, the South Protectorate (Cape Juby and Villa Bens/Tarfaya) and Western Sahara. Apart from the Spanish Army and the Spanish Navy, the EdA also took part in the conflict.

On 14 November 1957, Spanish Army officials met to decide on the measures that could be adopted to solve the crisis in Sidi Ifni. As a result, 24 CASA 2.111s, 15–20 HA-1112-M1Ls and 15–20 T-6s were sent to help in the conflict. As the fighting escalated, more aircraft were sent to the combat area, with a total of 100 Spanish combat aircraft (albeit mostly outdated) gathered in Sidi Ifni, the Canary Islands and Western Sahara on 4 February 1958. This included 30 CASA 2.111s, 30 Ju 52s, 12 T-6s and 14 HA-1112-M1Ls. Spain already had the F-86s, however, the conditions imposed by the US prohibiting the use of modern American-made jets meant that these aircraft could only be used on the Iberian

Pictured is a panel on the outside of the old Hispano Aviación factory in Seville, which was there between 1943 and 1972. Hispano Aviación manufactured various aircraft, such as the HA-1112-M1L and the fabulous Ha-200, the first jet manufactured in Spain, designed in 1955 by Willy Messerschmitt. (Eduardo Manuel Gil Martínez)

Peninsula. Officially, the Ifni War ended in May 1958, but in reality, the attacks on Spanish places and shootings against Spaniards continued during the following years, causing several deaths and injuries.

Between 1974 and 1975, F-5s from Ala 21 were deployed to bases in Gando, Canary Islands, and El Aaiun, Spanish Sahara, to protect Spanish air space from Moroccan aircraft. This time, as opposed to the Ifni War, there were no combat clashes, although it remained a real possibility. At the end of 1975, Operación *Golondrina* (Operation *Swallow*) occurred. Four DHC-4s from Ala 37 were used to retrieve Spanish personnel and goods from Spanish Sahara when Spain abandoned its claims in Western Africa. On 12 January 1976, Spanish aircraft flew for the last time over the Spanish Sahara, leaving it to become known as Western Sahara.

Between 1994 and 2002, the EdA deployed the Destacamento 'Ícaro' (Icarus Detachment) with eight EF-18s from Ala 12 and 15 Grupo, two KC-130s from Ala 31, one C-212 from Ala 37 and one P-3 from Ala 21 to Aviano AB and Sigonella AB, Italy, to assist in NATO operations *Deny Flight*, *Deliberate Force*, *Allied Force* and *Sharp Guard*. The first three operations were intended to control the air space over Bosnia, while *Sharp Guard* was designed to impose a naval embargo against the Yugoslavia Federal Republic. In these operations, the Spanish aircraft took part in combat missions with success.

During 2002, some F-18s and F-1s were deployed near the Gibraltar Strait to protect Spanish ground troops and Navy that had reconquered Perejil Island from Moroccan troops. However, in this case, combat flights did not take place.

The EdA is part of the NATO enhanced Air Policing (eAP) mission that presides over Estonia, Lithuania and Latvia. The Spanish Destacamento 'Ámbar' ('Amber' Detachment) carries out 'scramble'

As a homage to Spanish Aircraft Industries, the last picture in this book is a beautiful profile of C.4K-76 7-25 at Tablada AB, with many other HA-1112-M1Ls behind. Built in Sevilla, this aircraft came to life thanks to the combined efforts of the Spanish aeronautical industry, and it was in active service until 1965. It survived long enough to fly alongside modern jets, such as the F-104G.

missions intended to intercept enemy aircraft that do not announce their flight route. Six Spanish F-1s, EF-18s and EF2000s had been deployed to these countries.

In a similar way to the Baltic mission, the EdA is also part of another NATO surveillance mission in the Black Sea, where the EdA is still, at time of writing, taking part with six EF2000s from Ala 11.

The EdA has been part of Operation *Atalanta* since 2008. Spain's role here is to protect World Food Program ships from piracy off the coast of Somalia, using P-3s and CN-235s.

The vast majority of the international missions accomplished by the EdA have been providing humanitarian aid. In these missions, one aircraft stands out over the others –the wonderful C-130. Since 1975, this aircraft has flown in humanitarian aid missions in Niger, Mali, Indonesia, Philippines, Iraq, Afghanistan, Mexico, and Haiti, among others. Furthermore, EdA helicopters, such as the AS332, have intervened in international missions such as Operation *India-Mike* in Mozambique in 2000, where they provided air evacuations and delivered aid, and Afghanistan, where the aircraft were used in SAR and air evacuation missions. The last humanitarian mission featured A400Ms, which evacuated hundreds of persons, mainly civilians, from Afghanistan in late August 2021. Spanish aircraft have been deployed for aid and humanitarian missions all over the world with great success.

Bibliography

Arráez Cerdá, Juan, Photographic Files Collection.
Arráez Cerdá, Juan, 'Les Espagnols de la Luftwaffe, Les Escadrilles Bleues', *Ciel de Guerre Numero 18* (2010)
Arráez Cerdá, Juan, 'Les Espagnols de la Luftwaffe, Les Escadrilles Bleues', *Ciel de Guerre Numero 19* (2011)
Aviones militares españoles (1911–1986), Ministerio de Defensa (1986)
Caballero, Carlos and Guillén, Santiago, *Escuadrillas azules en Rusia: Historia y uniformes*, Almena Ediciones, Madrid (1999)
Gil Martínez, Eduardo Manuel, *Hispano Aviación HA-1112*, Kagero Publishing, Poland (2019)
Gil Martínez, Eduardo Manuel, *Spanish Air Force during World War II*, Kagero Publishing, Poland (2019)
González Serrano, José Luis, *Las unidades y el material del Ejército del Aire durante la Segunda Guerra mundial*, AF Editores, Spain (2005)
González Serrano, José Luis, *Recopilación números de tipo y denominaciones oficiales de los aviones del Ejército del Aire (1939–2012)*
Paloque, Gérard, *Avions de combat de L'Otan: Depuis 1949*, Heimdal Publishing (2020)
Pecker, Beatriz and Pérez, Carlos, *Crónica de la aviación española*, Silex (1983)
Sánchez Méndez, José, *La historia del helicóptero en el Ejército del Aire*, Eurocopter España (2007)
Full Collection, *Alas Españolas. Reserva anticipada ediciones*, (1999–2004)
'C-101, Eurofighter, F-1, F-5, F/A-18, Aviones de Transporte', *Avion Revue Internacional Magazine*, Key Publishing España (2019–21)